Introduction to Datafication

Implement Datafication Using AI and ML Algorithms

Shivakumar R. Goniwada

Apress®

Introduction to Datafication: Implement Datafication Using AI and ML Algorithms

Shivakumar R. Goniwada
Gubbalala, Bangalore, Karnataka, India

ISBN-13 (pbk): 978-1-4842-9495-6   ISBN-13 (electronic): 978-1-4842-9496-3
https://doi.org/10.1007/978-1-4842-9496-3

Managing Director, Apress Media LLC: Welmoed Spahr
Acquisitions Editor: Celestin Suresh John
Development Editor: Laura Berendson
Coordinating Editor: Mark Powers
Copy Editor: April Rondeau

Cover designed by eStudioCalamar

Cover image by Pawel Czerwinsk on Unsplash (www.unsplash.com)

Distributed to the book trade worldwide by Apress Media, LLC, 1 New York Plaza, New York, NY 10004, U.S.A. Phone 1-800-SPRINGER, fax (201) 348-4505, email orders-ny@springer-sbm.com, or visit www.springeronline.com. Apress Media, LLC is a California LLC and the sole member (owner) is Springer Science+Business Media Finance Inc. (SSBM Finance Inc.). SSBM Finance Inc. is a **Delaware** corporation.

For information on translations, please e-mail booktranslations@springernature.com; for reprint, paperback, or audio rights, please e-mail bookpermissions@springernature.com.

Apress titles may be purchased in bulk for academic, corporate, or promotional use. eBook versions and licenses are also available for most titles. For more information, reference our Print and eBook Bulk Sales web page at http://www.apress.com/bulk-sales.

Any source code or other supplementary material referenced by the author in this book is available to readers on GitHub (https://github.com/Apress). For more detailed information, please visit http://www.apress.com/source-code.

Printed on acid-free paper

This book is dedicated to those who may need access to the resources and opportunities many take for granted. May this book serve as a reminder that knowledge and learning are powerful tools that can transform lives and create new opportunities for those who seek them.

Table of Contents

About the Author

Shivakumar R. Goniwada is an author, inventor, chief enterprise architect, and technology leader with over 23 years of experience architecting cloud-native, data analytics, and event-driven systems. He works in Accenture and leads a highly experienced technology enterprise and cloud architect team. Over the years, he has led many complex projects across industries and the globe. He has ten software patents in cloud computing, polyglot architecture, software engineering, data analytics, and IoT. He authored a book on *Cloud Native Architecture and Design*. He is a speaker at multiple global and in-house conferences. Shivakumar has earned Master Technology Architecture, Google Professional, AWS, and data science certifications. He completed his executive MBA at the MIT Sloan School of Management.

About the Technical Reviewer

Dr. Mohan H M is a technical program manager and research engineer (HMI, AI/ML) at Digital Shark Technology, supporting the research and development of new products, promotion of existing products, and investigation of new applications for existing products.

In the past, he has worked as a technical education evangelist and has traveled extensively all over India delivering training on artificial intelligence, embedded systems, and Internet of Things (IoT) to research scholars and faculties in engineering colleges under the MeitY scheme. In the past, he has worked as an assistant professor at the T. John Institute of Technology. Mohan holds a master's degree in embedded systems and the VLSI design field from Visvesvaraya Technological University. He earned his Ph.D. on the topic of non-invasive myocardial infarction prediction using computational intelligence techniques from the same university. He has been a peer reviewer for technical publications, including BMC Informatics, Springer Nature, Scientific Reports, and more. His research interests include computer vision, IoT, and biomedical signal processing.

Acknowledgments

Many thanks to my mother, S. Jayamma, and late father, G.M. Rudrapp, who taught me the value of hard work, and to my wife, Nirmala, and daughter, Neeharika, without whom I wouldn't have been able to work long hours into the night every day of the week. Last but not least, I'd like to thank my friends, colleagues, and mentors at Mphasis, Accenture, and other corporations who have guided me throughout my career.

Thank you also to my colleagues Mark Powers, Celestin Suresh John, Shobana Srinivasan, and other Apress team members for allowing me to work with you and Apress, and to all who have helped this book become a reality. Thank you for my mentors Bert Hooyman and Abubacker Mohamed and thanks for my colleague Raghu Pasupuleti for providing key inputs.

Introduction

The motivation to write this book goes back to the words of Swami Vivekananda: "Everything is easy when you are busy, but nothing is easy when you are lazy," and "Take up on one idea, make that one idea your life, dream of it, think of it, live on that idea."

Data is increasingly shaping the world in which we live. The proliferation of digital devices, social media platforms, and the Internet of Things (IoT) has led to an explosion in the amount of data generated daily. This has created new opportunities and challenges for everyone as we seek to harness the power of data to drive innovation and improve decision making.

This book is a comprehensive guide to the world of datafication and its development, governing process, and security. We explore fundamental principles and patterns, analysis frameworks, techniques to implement artificial intelligence (AI) and machine learning (ML) algorithms, models, and regulations to govern datafication systems.

We will start by exploring the basics of datafication and how it transforms the world, and then delve into the fundamental principles and patterns and how data are ingested and processed with an extensive data analysis framework. We will examine the ethics, regulations, and security of datafication in a real scenario.

Throughout the book, we will use real-world examples and case studies to illustrate key concepts and techniques and provide practical guidance in sentiment and behavior analysis.

Whether you are a student, analyst, engineer, technologist, or someone simply interested in the world of datafication, this book will provide you with a comprehensive understanding of datafication.

## CHAPTER 1

Introduction to Datafication

A comprehensive look at datafication must first begin with its definition. This chapter provides that and details why datafication plays a significant role in modern business and data architecture.

Datafication has profoundly impacted many aspects of society, including business, finance, health care, politics, and education. It has enabled companies to gain insights into consumer behavior and preferences, health care to improve patient outcomes, finance to enhance consumer experience and risk and compliance, and educators to personalize learning experiences.

Datafication helps you to take facts and statistics gained from myriad sources and give them domain-specific context, aggregating and making them accessible for use in strategy building and decision making. This improves sales and profiles, health results, and influence over public policy.

Datafication is the process of turning data into a usable and accessible format and involves the following:

- Collecting data from myriad sources

- Organizing and cleaning the data

© Shivakumar R. Goniwada 2023
S. R. Goniwada, *Introduction to Datafication*,
https://doi.org/10.1007/978-1-4842-9496-3_1

- Making it available for analysis to use

- Analyzing the data by using artificial intelligence (AL) and machine learning (ML) models

Developing a deeper understanding of the datafication process and its implications for individuals and society is essential. This requires a multidisciplinary approach that brings together stakeholders from various fields to explore the challenges and opportunities of datafication and to develop ethical and effective strategies for managing and utilizing data in the digital age.

This chapter will drill down into the particulars and explain how datafication benefits the across industry. We will cover the following topics:

- What is datafication?

- How is datafication embraced across industries?

- Why is datafication important?

- What are elements of datafication?

What Is Datafication?

Datafication involves using digital technologies such as the cloud, data products, and AI/ML algorithms to collect and process vast amounts of data on human behavior, preferences, and activities.

Datafication converts various forms of information, such as texts, images, audio recordings, comments, claps, and likes/dislikes to curated format, and that data can be easily analyzed and processed by multiple algorithms. This involves extracting relevant data from social media, hospitals, and Internet of Things (IoT). These data are organized into a consistent format and stored in a way that makes them accessible for further analysis.

Everything around us, from finance, medical, construction, and social media to industrial equipment, is converted into data. For example, you create data every time you post to social media platforms such as WhatsApp, Instagram, Twitter, or Facebook, and any time you join meetings in Zoom or Google Meet, or even when you walk past a CCTV camera while crossing the street. The notion differs from digitization, as datafication is much broader than digitization.

Datafication can help you to understand the world more fully than ever before. New cloud technologies are available to ingest, store, process, and analyze data. For example, marketing companies use Facebook and Twitter data to determine and predict sales. Digital Twin uses industrial equipment behavior to analyze the behavior of the machine.

Datafication also raises important questions about privacy, security, and ethics. The collection and use of personal data can infringe on individual rights and privacy, and there is a need for greater transparency and accountability in how data are collected and used. Overall, datafication represents a significant shift in how we live, work, and act.

Why Is Datafication Important?

Datafication enables organizations to transform raw data into a format that can be analyzed and used to gain insights, make informed business decisions, improve patients' health, and streamline supply-chain management. This is crucial for every industry to improve in today's data-driven world. By using the processed data, organizations can identify trends, gain insight into customer behavior, and discover other key performance indicators using analytics tools and algorithms.

Data for Datafication

Data is available everywhere, but what *type* of data you require for analysis in datafication is crucial and helps you to understand hidden values and challenges. Data can come from a wide range of sources, but the specific data set will depend on the particular context and the goal of the datafication process.

Today, data are created not only by people and their activities in the world, but also by machines. The amount of data produced is almost out of control.

For example:

- Social media data such as posts and comments are structured data that can be easily analyzed for sentiment and behavior. This involves extracting text from the posts and comments and identifying and categorizing any images, comments, or other media that are part of it.

- In the medical context, datafication might involve converting medical records and other patient information into structured data that can be used for analysis and research. This involves extracting information about diagnoses, treatments, and other medical reports.

- In the e-commerce context, datafication might involve converting users' statistics and other purchase information into structured data that can be used for analysis and recommendations.

In summary, data can come from a wide range of sources, and how it is used will depend on the specific context and goals of the datafication process.

Data constantly poses new challenges in terms of storage and accessibility. The need to use all of this data is pushing us into a higher level of technological advancement, whether we like or want it or not.

Datafication requires new forms of integration to uncover large hidden values from extensive collections that are diverse, complex, and of a massive scale. According to Kepios (`https://kepios.com/`), there will be 4.80 billion social media users worldwide as of April 2023, 59.0 percent of the world population, and approximately 227 million users join every year.

The following are a few statistics regarding major social media applications as of the writing of this book:

- Facebook has 3.46 billion monthly visitors.

- YouTube's potential advertising reach is 7.55 billion people (monthly average).

- WhatsApp has at least 3 billion monthly users.

- Instagram's potential advertising reach is approximately 2.13 billion people.

- Twitter's possible advertising reach is approximately 2.30 billion people.

Datafication Steps

For datafication, as defined by DAMA (Data Management Association), you must have a clear set of data, well-defined analysis models, and computing power. To obtain a precise collection of data, relevant models, and required computing power, one must follow these steps:

- **Data Harvesting**: This step involves obtaining data in a real-time and reliable way from various sources, such as databases, sensors, files, etc.

- **Data Curation**: This step involves organizing and cleaning the data to prepare it for analysis. You need to ensure that the data collected are accurate by removing errors, inconsistencies, and duplicates with a standardized format.

- **Data Transformation**: This step involves converting data into a suitable format for analysis. This step helps you transform the data into a specific form, such as dimensional and graph models.

- **Data Storage**: This step involves storing the data after transformation in storage, such as a data lake or data warehouse, for further analysis.

- **Data Analysis**: This step involves using statistical and analytical techniques to gain insights from data and identify trends, patterns, and correlations in the data that help with predictions and recommendations.

- **Data Dissemination**: This step involves sharing the dashboards, reports, and presentations with relevant stakeholders.

- **Cloud Computing**: This step provides the necessary infrastructure and tools for the preceding steps.

Digitization vs. Datafication

For a better understanding of datafication, it can be helpful to contrast it with digitization. This may help you to better visualize the datafication process.

Digitization is a process that has taken place for decades. It entails the conversion of information into a digital format; for example, music to MP3/MP4, images to JPG/PNG, manual banking process to mobile and automated web process, manual approval process to automatic BPM workflow process, and so on.

Datafication, on the other hand, involves converting data into a usable, accessible format. This consists of collecting data from various sources and formats, organizing and cleansing it, and making it available for analysis. The primary goal of datafication is to help the organization make data-driven decisions, allowing it to gain insights and knowledge from the data.

Datafication helps monitor what each person does. It does so with advanced technologies that can monitor and measure things individually.

In digitization, you convert many forms into digital forms, which are accessible to an individual computer. Similar to datafication, you ingest the activities and behavior and convert them into a virtual structure that can be used within formal systems.

However, many organizations realize that more than simply processing data is needed to support business disruption. It requires quality data and the application of suitable algorithms. Modern architecture and methodologies must be adopted to address these challenges to create datafication opportunities.

Types of Data in Datafication

The first type of data is content, which can be user likes, comments on blogs and web forums, visible hyperlinks in the content, user profiles on social networking sites, news articles on news sites, and machine data. The data format can be structured or unstructured.

The second type of data is the behavior of objects and the runtime operational parameters of industrial systems, buildings, and so forth.

The third type is time series data, such as stock price, weather, or sensor data.

The fourth type of data is network structured data, such as integrated networked systems in an industrial unit, such as coolant pipes and water flow. This data type is beneficial because it provides for overall media analysis, entire industrial function, and so on.

The fifth data set is your health, fitness, sleep time, conversation chats, smart home, and health monitor device.

Elements of Datafication

As defined by DAMA, Figure 1-1 illustrates the seven critical elements of the datafication architecture used to develop the datafication process. Datafication will only be successful if at least one of the steps is included.

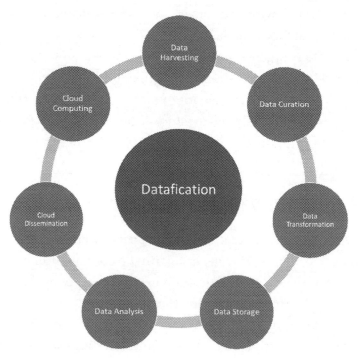

Figure 1-1. *Data elements*

Data Harvesting

Data harvesting is extracting data from a given source, such as social media, IoT devices, or other various data sources.

Before harvesting any data, you need to analyze it to identify the source and software tools needed for harvesting.

First, the data is undesiably noticeable if it is inaccurate, biased, confidential, and irrelevant. Therefore, harvested information is more objective and reliable than familiar data sources. However, the disadvantage is that it is difficult to know the users' demographic and psychological variables for social media data sources.

Second, harvesting must be automatic, real-time, streaming and able to handle large-scale data sources efficiently.

Third, the data are usually fine-grained and available in real-time. Text mining techniques are used to preprocess raw text images, text processing techniques are used to preprocess essential texts, and video processing techniques are used to preprocess photos and videos for further analysis.

Fourth, the data can be ingested in real-time or in batches. In real-time, each data item is imported as the source changes it. When data are ingested through sets, the data elements are imported in discrete chunks at periodic intervals.

Various data harvesting methods can be used depending on the data source type, as follows:

- IoT devices typically involve collecting data from IoT sensors and devices using protocols such as MQTT, CoAP, HTTP, and AMQP.

- Social media platforms such as Facebook, Twitter, LinkedIn, Instagram, and others use REST API, streaming, Webhooks, and GraphQL.

Data Curation

Data curation organizes and manages data collected through ingestion from various sources. This involves organizing and maintaining data in a way that makes it accessible and usable for data analysis. This involves cleaning and filtering data, removing duplicates and errors, and properly labeling and annotating data.

Data curation is essential for ensuring that the data are accurate, consistent, and reliable, which is crucial for data analysis.

The following are the few steps involved in data curation:

- **Data Cleaning**: Once the data is harvested, it must be cleaned to remove errors, inconsistencies, and duplicates. This involves removing missing values, correcting spelling errors, and standardizing data formats.

- **Data Transformation**: After the data has been cleaned, it needs to be transformed into a format suitable for analysis. This involves aggregating data, creating new variables, and so forth. For example, you might have a data set of pathology reports with variables that include such elements as patient ID, date of visit, test ID, test description, and test results. You want to transform this data set into a format that shows each patient's total health condition. To do this you need to alter harvested data for data analysis with transformations such as creating a new variable for test category, aggregating test data for patient for a year, summarizing data by group (ex: hemoglobin), etc.

- **Data Labeling**: Annotating data with relevant metadata, such as variable names and data descriptions.

- **Data Quality Test**: In this step, you need to ensure the data is accurate and reliable by using various tests like statistical tests, etc.

The overall objective of data curation is to reduce the time it takes to obtain insight from raw data by organizing and bringing relevant information together for further analysis.

The steps involved in data curation are organizing and cataloging data, ensuring data quality, preserving data, and providing access to data.

Data Storage

Data storage stores actual digital data on a computer with the help of a hard drive, solid-state drive, and related software to manage and organize the data.

Data storage is the actual physical storage of datafication data. More than 2.5 quintillion bytes of data are created daily, and data snowballs of approximately 2 MB are made every second for every person. These numbers are from users searching the content in the internet, browsing social media networks, posting blogs, photos, comments, status updates, watching a video, downloading images, streaming songs, etc. To make a business decision, the data must be stored in a way that is easier to manage and access, and it is essential to protect data against cyber threats.

For IoT, the data need to be collected from sensors and devices and stored in the cloud.

Several types of database storage exist, including relational databases, NoSQL databases, in-memory databases, and cloud databases. Each type of database storage has advantages and disadvantages, but the best choice for datafication is cloud databases that involve data lakes and warehouses.

Data Analysis

Data analysis refers to analyzing a large set of data to discover different patterns and KPIs (Key Performance Indicators) of an organization. The main goal of analytics is to help organizations make better business decisions and future predictions. Advanced analytics techniques such as machine learning models, text analytics, predictive analytics, data mining, statistics, and natural language processing are used. With these ML models, you uncover hidden patterns, unknown correlations, market trends, customer preferences, feedback about your new FMCG (Fast Moving Consumer Goods) products, and so on.

The following are the types of analytics that you can process using ML models:

- **Prescriptive**: This type of analytics helps to decide what action should be taken and examines data to answer various questions such as what should be done. Or what can we do to make our product attractive? This helps to find an answer to various problems, such as where to focus on treatment.

- **Predictive**: This type of analytics helps to predict the future or what might happen, such as emphasizing the business relevance of the resulting insights and use cases, such as sales and production data.

- **Diagnostic**: This type of analytics helps to analyze past situations, such as what went wrong and why it happened. This helps to facilitate correction in the future; for example, weather prediction and customer behavior.

- **Descriptive**: This type of analytics helps to analyze current and future use cases, such as behavioral analysis of users.

- **Exploratory**: This type of analytics involves visualization.

Cloud Computing

Cloud computing is the use of computing resources delivered over the internet and has the potential to offer substantial opportunities in various datafication scenarios. It is a flexible delivery platform for data, computing, and other services. It can support many architectural and development styles, from extensive, monolithic systems to sizeable virtual machine deployments, nimble clusters of containers, a data mesh, and large farms of serverless functions.

The primary services of cloud offerings for data storage are as follows:

- Data storage as a service

- Streaming services for data ingestion

- Machine learning workbench for analysis

Datafication Across Industries

Datafication is a valuable resource for businesses and organizations seeking to gain insights into customer behavior, market trends, patient health healing progress, and more.

Datafication is the process of converting various types of data into a standardized format that can be used for analysis and decision making and has become increasingly important across industries as a means of leveraging data.

In the health-care industry, datafication is used to improve patient outcomes and reduce costs. By collecting and analyzing patient data, including pathology tests, medical histories, vital signs, and lab results, health-care providers are able to optimize treatments and improve patient care.

In the finance industry, datafication is used to analyze financial data, such as transaction history, risk, fraud management, personalized customer experience, and compliance.

In the manufacturing industry, datafication is used to analyze production data, machine data to improve the production process, digital twins, etc.

In the retail industry, datafication is used to analyze customer behavior and preferences to optimize pricing strategies and personalized customer experience.

Summary

Datafication is the process of converting various types of data and information into a digital format that can easily be processed and analyzed. With datafication, you can increase your organization's footprint by using data effectively for decision making. It helps to improve operational efficiency and provides input to the manufacturing hub to develop new products and services.

Overall, data curation is the key component of effective datafication, as it ensures that the data is accurate, complete, and reliable, which is essential for making decisions and gleaning meaningful insights.

In this chapter, I described datafication and discussed the types of data involved in datafication, datafication steps, and datafication elements. Next chapter provides more details of principles, patterns and methodolgoies to realize the datafication.

CHAPTER 2

Datafication Principles and Patterns

Principles are guidelines for the design and development of a system. They reflect the level of consensus among the various elements of your system. Without proper principles, your architecture has no compass to guide its journey toward datafication.

Patterns are tried-and-tested accurate solutions to common design problems, and they can be used as a starting point for developing a datafication.

The processes involved in datafication are to collect, analyze, and interpret the vast amount of information from a range of sources, such as social media, Internet of Things (IoT) sensors, and other devices. The principles and patterns underlying datafication must be understood to ensure that it benefits all.

The patterns are reusable solutions to commonly occurring problems in software design. These patterns provide a template for creating designs that solve specific problems while also being flexible to adapt to different contexts and requirements.

© Shivakumar R. Goniwada 2023
S. R. Goniwada, *Introduction to Datafication*,
https://doi.org/10.1007/978-1-4842-9496-3_2

This chapter provides you with an overview of the principles and patterns shaping the development of datafication. It will examine the ethical implication of these technologies for society. Using these principles and patterns, you can develop datafication projects that are fair and transparent and that perform well.

What Are Architecture Principles?

A principle is a law or a rule that must be or usually is to be followed when making critical architectural decisions. The architecture and design principles of datafication play a crucial role in guiding the software architecture work responsible for defining the datafication direction. While following these principles, you must also align with the existing enterprise's regulations, primarily those related to data and analytics.

The data and analytics architecture principles are a subset of the overall enterprise architecture principles that pertain to the rules surrounding your data collection, usage, management, integration, and analytics. Ultimately, these principles keep your datafication architecture consistent, clean, and accountable and help to improve your overall datafication strategy.

Datafication Principles

As mentioned in the previous chapter, datafication analyzes data from various sources, such as social media, IoT, and other digital devices. For a successful and streamlined datafication architecture, you must define principles related to data ingestion, data streaming, data quality, data governance, data storage, data analysis, visualization, and metrics. These principles ensure that data and analytics are used in a way that is aligned with the organization's goals and objectives.

Examples of datafication principles include the use of accurate and up-to-date data, the use of a governance framework, the application of ethical standards, and the application of quality rules.

The following few principles that helps you to design the datafication process:

Data Integration Principle

Before big data and streaming technology, data movement was simple. Data moved linearly from static structured databases and static APIs to data warehouses. Once you built an integration pipeline in this stagnant world, it operated consistently because data moved like trains on a track.

In datafication, data have upended the traditional train track–based approach to use a modern and smart city traffic signal–based approach. To move data at the speed of business and unlock the flexibility of modern data architecture, the integration must be handled such that it has the ability to monitor and manage performance continually. For modern data integration, your data movement must be prepared for the following:[1]

- Be capable of doing streaming, batch, API-led, and micro-batch processing

- Support structured, semi-structured, and unstructured data

- Handle scheme and semantic changes without affecting the downstream analysis

- Respond to changes from sources and application control

---

[1] *https://streamsets.com/blog/data-integration-architecture/*

The following principles will help you design modern data integration. For example, in health-care data analysis, you need to integrate various health-care systems in the hospitals, such as electronic medical records and insurance claims data. In financial data analysis, to generate trends and statistics of financial performance, you need to integrate various data systems, such as payment processors, accounting systems, and so forth.

- **Design for Both Batch and Streaming**: While you are building for social media and IoT, which capitalize on streaming and API-led data, you must account for the fact that these data often need to be joined with or analyzed against historical data and other batch sources within an enterprise.

- **Structured, Semi-structured, and Unstructured Data**: Data integration combines data from multiple sources to provide a unified view. To achieve this, data integration software must be able to support this.

- **Handle Scheme and Schematics Changes**: In data integration, it is standard for the scheme and schematics of the data to change over time as new data sources are added or existing sources are modified. These changes affect the downstream analysis, making it difficult to maintain the data's integrity and the analysis's accuracy. It is essential to use a flexible and extensible data integration architecture to handle this. You can use data lineage tools to achieve this.

- **Respond to Changes from Sources**: In data integration, responding to the source side requires technical and organizational maturity. Using CDC (Change Data Capture) and APIs (Application Programming Interface) and implementing the best change management ensures that data integration is responsive, efficient, and effective.

- **Use Low-Code No-Code Concepts**: Writing custom code to ingest data from the source into your data store has been commonplace.

- **Sanitize Raw Data upon Data Harvest**: Storing raw inputs invariably leads you to have personal data and otherwise sensitive information posing some compliance risks (use only when it is needed). Sanitizing data as close to the source as possible makes data analytics productive.

- **Handle Data Drift to Ensure Consumption-Ready Data**: Data drift refers to the process of data changing over time, often in ways that are unpredictable and difficult to detect. This drift can occur for many reasons, such as changes in the data source, changes in data processing algorithms, or changes in the system's state. This kind of drift can impact the quality and reliability of data and analytics. Data drift increases costs, causes delays in time to analysis, and leads to poor decisions based on incomplete data. To mitigate this, you need to analyze and choose the right tools and software to detect and react to changes in the schema and keep data sources in sync.

- **Cloud Design**: Designing integration for the cloud is fundamentally different when architecting the cloud. Enterprises often put raw data into object stores without knowing the end analytical intent. Legacy tools for data integration often lack the level of customization and interoperability needed to take full advantage of cloud services.

- **Instrument Everything**: Obtaining end-to-end insight into data systems will be challenging. End-to-end instrumentation helps to manage data movements. This instrumentation is needed for time series analysis of a single data flow to tease out changes over time.

- **Implement the DataOps Approach**: Traditional data integration was suitable for the waterfall delivery approach but may not work for modern-day engineering principles. Modern dataflow tools provide an integrated development environment for continuous use for the dataflow life cycle.

Data Quality Principle

Ensuring you have high-quality data is central to the data management platform. The principle of data quality management is a set of fundamental understandings, standards, rules, and values. It is the core ingredient of a robust data architecture. Data quality is critical for building an effective datafication architecture. Well-governed, high-quality data helps create accurate models and robust schemas.

There are five characteristics of data quality, as follows:

- **Accuracy**: Is the information captured in every detail?

- **Completeness**: How compressive is the data?

- **Reliability**: Does the data contradict other trusted resources?

- **Relevance**: Do you need this data?

- **Timeliness**: Is this data obsolete or up-to-date, and can it be used for real-time analysis?

To address these issues, several steps can be taken to improve the quality, as follows:

- Identify the type of source and assess its reliability.

- Check the incoming data for errors, inconsistencies, and missing values.

- Use data cleaning techniques to fix any issues and improve the quality.

- Use validation rules to ensure data is accurate and complete. This could be an iterative approach.

- Monitor regularly and identify changes.

- Apply data governance and management process.

Data Quality Tools

Data quality is a critical capability of datafication, as the accuracy and reliability of data are essential for an accurate outcome. These tools and techniques can ensure that data is correct, complete, and consistent and can help identify and remediate quality issues. There are various tools and techniques to address data quality. Here are a few examples:

- **Data Cleansing tools**: These help you identify and fix errors, inconsistencies, and missing values.

- **Data Validation tools**: These tools help you to check data consistency and accuracy.

- **Data Profiling tools**: These will provide detailed data analysis such as data types, patterns, and trends.

- **Data Cataloging tools**: These tools will create a centralized metadata repository, including data quality metrics, data lineage, and data relationships.

- **Data Monitoring and Alerting tools**: These track data quality metrics and alert the governance team when quality issues arise.

Data Governance Principles

Data are an increasingly significant asset as organizations implementing datafication move up the digital curve as they focus on big data and analytics. Data governance helps organizations better manage data availability, usability, integrity, and security. It involves establishing policies and procedures for collecting, storing, and using data.

In modern architecture, especially for datafication, data are expected to be harvested and accessed anywhere, anytime, on any device. Satisfying these expectations can give rise to considerable security and compliance risks, so robust data governance is needed to meet the datafication process.

Data governance is about bringing data under control and keeping it secure. Successful data governance requires understanding the data, policies, and quality of metadata management, as well as knowing where data resides. How did it originate? Who has access to it? And what does it mean? Effective data governance is a prerequisite to maintaining business compliance, regardless of whether that compliance is self-imposed by an organization or comes from global industry practices.

Data governance includes how data delivery and access take place. How is data integrity managed? How does data lineage take place? How is data loss prevention (DLP) configured? How is security implemented for data?

Data governance typically involves the following:

- Establish the data governance team and define its roles and responsibilities.

- Develop a data governance framework that includes policies, standards, and procedures.

- Data consistency across user behavior ensures completeness and accuracy in generating required KPIs (Key Performance Indicators).

- Identify critical data assets and classify them according to their importance.

- Define compliance matrices like GDPR, etc.

- Fact-based decisions based on advanced analytics become actual time events, and data governance ensures data veracity, which builds the confidence an organization needs to achieve the real-time goal for decision making.

- Consider using data governance software such as Alation, Collibra, Informatica, etc.

Data Is an Asset

Data is an asset that has value to organizations and must be managed accordingly. Data is an organizational resource with real measurable value, informing decisions, improving operations, and driving business growth. Organizations' assets are carefully managed, and data are equally important as physical or digital assets. Quality data are the foundation of the organization's decisions, so you must ensure that the data are harvested with quality and accuracy and are available when needed. The techniques used to measure the data value are directly related to the accuracy of the outcome of the decision, the accuracy of the outcome depends on the quality, relevance, and reliability of hte data used in the decision-making process. the common techniques are data quality assessment, data relevance analysis, cost-benefit analysis, impact analysis and differnet forms of analytics.

Data Is Shared

Different organizational stakeholders will access the datafication data to analyze various KPIs. Therefore, the data can be shared with relevant teams across an organization. Timely access to accurate and cleansed data is essential to improving the quality and efficiency of an organization's decision-making ability. The speed of data collection, creation, transfer, and assimilation is driven by the ability of an organization's process and technology to capture social media or IoT sensor data.

To enable data sharing, you must develop and abide by a common set of policies, procedures, and standards governing data management and access in the short and long term. It would be best if you had a clear blueprint for data sharing; there should not be any compromise of the confidentiality and privacy of data.

Data Trustee

Each data element in a datafication architecture has a trustee accountable for its quality. As the degree of data sharing grows and business units within an organization rely upon information, it becomes essential that only the data trustee makes decisions about the content of the data. In this role, the data trustee is responsible for ensuring that the data used are following applicable laws, regulations, or policies and are handled securely and responsibly. The specific responsibilities of a data trustee will vary depending on the type of data being shared and the context in which it is being used.

The trustee and steward are different roles. The trustee is responsible for the accuracy and currency of the data, while the steward may be broader and include standardization and definition tasks.

Ethical Principle

Datafication focuses on and analyzes social media, medical, and IoT data. These data are focused on human dignity, which involves considering the potential consequences of data and ensuring that it is used fairly, responsibly, and transparently. This principle reflects the fundamental ethical requirement that people be treated in a way that respects their dignity and autonomy as human individuals. When analyzing social media and medical data, we must remember that data also affects, represents, and touches people. Personal data are entirely different from any machine's raw data, and the unethical use of personal data can directly influence people's interactions, places in the community, personal product usage, etc. It would be best if you considered various laws across the globe to meet ethics needs while designing your system.

There are various laws in place globally; here are a few:

GDPR Principles (Privacy): Its focus is protecting, collecting, and managing personal data; i.e., data about individuals. It applies to all companies and originations in the EU and companies outside of Europe that hold or otherwise process personal data. The following are a few guidelines from the GDPR. For more details, refer to `https://gdpr-info.eu/`:

- **Fairness, Lawfulness, Transparency**: Personal data shall be processed lawfully, fairly, and transparently about the data subject.

- **Purpose Limitation**: Personal data must be collected for specified, explicit, and legitimate purposes and not processed in an incompatible manner.

- **Data Minimization**: Personal data must be adequate, relevant, and limited to what is necessary for the purpose they are processed.

- **Accuracy**: Personal data must be accurate and, where necessary, kept up to date.

- **Integrity and Confidentiality**: Data must be processed with appropriate security of the personal data, including protection against unauthorized and unlawful processing.

- **Accountability**: Data controllers must be responsible for any compliance

PIPEDA (Personal Information Protection and Electronic Documents Act): This applies to every organization that collects, uses, and disseminates personal information. The following are the statutory obligations of PIPEDA; for more information, visit https://www.priv.gc.ca/:

- **Accountability**: Organizations are responsible for personal information under its control and must designate an individual accountable for compliance.

- **Identifying Purpose**: You must specify the purpose for which personal information is collected.

- **Consent**: You must obtain the knowledge and consent of the individual for the collection.

- **Accuracy**: Personal information must be accurate, complete, and up to date.

- **Safeguards**: You must protect personal information.

Human Rights and Technology Act: The U.K. government proposed this act. It would require companies to conduct due diligence to ensure that their datafication system does not violate human rights and to report any risk or harm associated with the technology. You can find more information at https://www.equalityhumanrights.com/. The following are a few guidelines:

- **Human Rights Impact Assessment**: Conduct a human rights impact assessment before launching new services.

- **Transparency and Accountability**: You must disclose information about technology services, including how you collect the data and the algorithms you use to make decisions affecting individual rights.

Universal Guidelines for AI: This law provides a set of guidelines for AI/ML and was developed by IEEE (Institute of Electrical and Electronics Engineers). These guidelines include transparency, accountability, and safety. You can find more information at `https://thepublicvoice.org/ai-universal-guidelines/`. The following are a few guidelines:

- **Transparency**: AI should be transparent in decision-making process, and data algorithms used in AI should be open and explainable.

- **Safety and Well-being**: Should be designed to ensure the safety and well-being of individuals and society.

There are various laws available for each country, and we suggest following the laws and compliance requirements before processing any data for analysis.

Security by Design Principle

Security by design also means privacy by design and is a concept in which security and privacy are considered fundamental aspects of the design.

This principle emphasizes the importance of keeping security and privacy at the core of a product system.

The following practices help with the design and development of a datafication architecture:[2]

- **Minimize Attack Surface Area**: Restricts a user's access to services.

- **Establish Secure Defaults**: Strong security rules on registering users to access your services.

- **The Principle of Least Privilege**: The user should have minimum privileges needed to perform a particular task.

- **The Principle of Defense Depth**: Add multiple layers of security validations.

- **Fail Securely**: Failure is unavoidable and therefore you want it to fail securely.

- **Don't Trust Services**: Do not trust third-party services without implementing a security mechanism.

- **Separation of Duties**: Prevent individuals from acting fraudulently.

- **Avoid Security by Obscurity**: Should be sufficient security controls in place to keep your application safe without hiding core functionality or source code.

- **Keep Security Simple**: Avoid the use of very sophisticated architecture when developing security controls.

- **Fix Security Issues Correctly**: Developers should carefully identify all affected systems.

---

[2] Cloud Native Architecture and Design Patterns, *Shivakumar Goniwada, Apress, 2021*

Datafication Patterns

Datafication is the process of converting various aspects of invisible data into digital data that can be analyzed and used for decision making. As I explained in Chapter 1, "Introduction to Datafication," datafication is increasingly prevalent in recent years, as advances in technology have made it easier to collect, store, and analyze large amount of data.

The datafication patterns are the common approaches and techniques used in the process of datafication. These patterns involve the use of various technologies and methods, digitization, aggregation, visualization, AI, and ML to convert data into useful insights.

By understanding these patterns, you can effectively store, collect, analyze, and use data to drive decision making and gain a competitive edge. By leveraging these patterns, you optimize storage operations.

Each solution is stated so that it gives the essential fields of the relationships needed to solve the problem, but in a very general and abstract way so that you can solve the problem for yourself by adapting it to your preferences and conditions.

The patterns can be the following:

- Can be seen as building blocks of more complex solutions

- The function is a common language used by technology architects and designers to describe solutions.[3]

---

[3] Cloud Native Architecture and Design Patterns, *Shivakumar Goniwada, Apress, 2021*

Data Partitioning Pattern

Partition allows a table, index, or index-organized table to be subdivided into smaller chunks, where each chunk of such a database object is called a partition. This is often done for reasons of efficiency, scalability, or security.

Data partitioning divides the data set and distributes the data over multiple servers or shards. Each shard is an independent database, and collectively the shards make up a single database. The portioning helps with manageability, performance, high availability, security, operational flexibility, and scalability.

Data partitioning addresses the following scale-like issues:

- High query rates exhausting the CPU capacity of the server

- Larger data sets exceeding the storage capacity of a single machine

- Working set sizes are more significant than the system's RAM, thus stressing the I/O capacity of disk drives.

You can use the following strategies for database partitioning:

- **Horizontal Partitioning (Sharding)**: Each partition is a separate data store, but all partitions have the same schema. Each partition is known as a shard and holds a subset of data.

- **Vertical Partitioning**: Each partition holds a subset of the fields for items in the data store. These fields are divided according to how you access the data.

- **Functional Partitioning**: Data are aggregated according to how each bounded context in the system uses it.

You can combine multiple strategies in your application. For example, you can apply horizontal partitioning for high availability and use a vertical partitioning strategy to store based on data access.

The database, either RDBMS or NoSQL, provides different criteria to share the database. These criteria are as follows:

- Range or interval partitioning

- List partitioning

- Round-robin partitioning

- Hash partitioning

Round-robin partitioning is a data partitioning strategy used in distributed computing systems. In this strategy, data is divided into equal-sized partitions or chunks and assigned to different nodes in a round-robin fashion. It distributes the rows of a table among the nodes. In range, list, and hash partitioning, an attribute "partitioning key" must be chosen from among the table attributes. The partition of the table rows is based on the value of the partitioning key. For example, if there are three nodes and 150 records to be partitioned, the records are divided into three equal chunks of 50 records each. The first chunk is assigned to the first node, the second chunk assigned to the second node, and so on. After each node is assigned a chunk of data, the partitioning starts again from the beginning, assigning the fifth chunk to the first node and so on.

Range partitioning is a partitioning strategy where data is partitioned based on a specific range of values. For example, you have a large data set of patient records, and you want to partition the data based on the age group. To do this, first you determine the minimum and maximum age groups in the data set and then divide the range of dates into equal intervals, each representing a partition.

Data Replication

Data replication is the process of copying data from one location to another location. The two locations are generally located on different servers. This kind of distribution satisfies the failover and fault tolerance characteristics.

Replication can serve many nonfunctional requirements, such as the following:

- **Scalability**: Can handle higher query throughput than a single machine can handle

- **High Availability**: Keeping the system running even when one or more nodes go down

- **Disconnected Operations**: Allowing an application to continue working when there is a network problem

- **Latency**: Placing data geographically closer to users so that users can interact with the data faster

In some cases, replication can provide increased read capacity as the client can send read operations to different servers. Maintaining copies of data in different nodes and different availability zones can increase the data locality and availability of the distributed application. You can also maintain additional copies of dedicated purposes, such as disaster recovery, reporting, or backup.

There are two types of replication:

- Leader-based or leader-follower replication

- Quorum-based replication

These two types of replication support full data replication, partial data replication, master-slave replication, and multi-master replication.[4]

---

[4] Cloud Native Architecture and Design Patterns, *Shivakumar Goniwada, Apress, 2021*

Stream Processing

Stream processing is the real-time processing of data streams. A stream is a continuous flow of data that is generated by a variety of sources, such as social media, medical data, sensors, and financial transactions.

Stream processing helps consumers query continuous data streams to detect conditions (for example, in payment processing, the AML (Anti-Money Laundering) system alerts if it founds anamolies in transactions) quickly in a near-real-time mode instead of batch mode. The detection of the condition varies depending on the type of source and use cases used.

There are several approaches to stream processing, including stream processing application frameworks, application engines, and platforms. Stream processing allows applications to exploit a limited form of parallel processing more easily. The application that supports stream processing can manage multiple computational units without explicitly managing allocation, synchronization, or communication among those units. The stream processing pattern simplifies parallel software and hardware by restricting the parallel computations that can be performed.

Stream processing takes on data via aggregation, analytics, transformations, enrichment, and ingestion.

As shown in Figure 2-1, for each input source, the stream processing engine operates in real time on the data source and provides output in the target database.

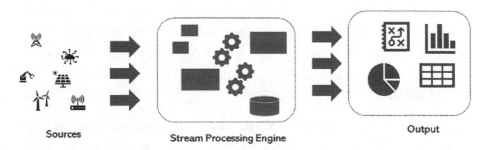

Sources   Stream Processing Engine   Output

Figure 2-1. *Stream processing*

The output is delivered to a streaming analytics application and added to the output streams.

The stream processing pattern addresses many challenges in the modern architecture of real-time analytics and event-driven applications, such as the following:

- Stream processing can handle data volumes that are much larger than the data processing systems.

- Stream processing easily models the continuous flow of data.

- Stream processing decentralizes and decouples the infrastructure.

The typical use cases of stream processing will be examined next.

Social Media Data Use Case

Let's consider a real-time sentiment analysis. Sentiment analysis is the process of analyzing text data to determine the attitude expressed in the text, video, etc. Let's consider an e-commerce platform. They sell smart phones, and the company wants to monitor public opinion about different smart phone brands on social media, and to respond quickly to any negative feedback or complaints. To do this, you need to set up a stream processing to continuously monitor social media platforms for the mention of the various brands. The system can use natural language processing (NLP) technique to perform sentiment analysis on the text of the posts and classifies each mention as positive, negative and neutral.

IoT Sensors

Stream processing is used for real-time analysis of data generated by IoT sensors. Let's consider a boiler machine at a chemical plant. They have a network of IoT sensors that are used to monitor environmental condition, temperature, humidity, etc. The company wants to use the data generated by the boiler to optimize the chemical process and detect potential issues in real-time. To do this, you need stream processing to continuously analyze the sensor data. The system uses ML algorithms to detect anomalies or patterns in the data and triggers alerts when certain thresholds are met.

Geospatial Data Processing

Stream processing is used for real-time analysis of data generated by GPS tracking. Let's consider an example with a shipping company. It wants to optimize its fleet management operation by tracking the location and status of each container in real-time. You can use GPS tracking on the containers to collect location data, which is then streamed to a stream processing system for real-time analysis.

Change Data Capture (CDC)

CDC is a replication solution that captures database changes as they happen and delivers them to the target database. CDC is a modern cloud architecture, as this pattern is a highly efficient way to move data from the source to the target, which helps to generate real-time analytics. CDC is a part of the ETL (extract, transform, and load) process, where data are extracted from the source, transformed, and loaded into target resources, such as a data lake or a data warehouse, as shown in Figure 2-2.

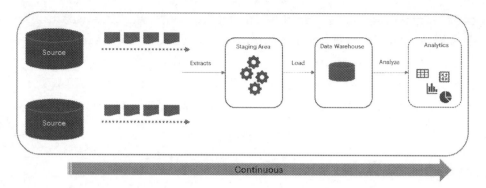

Figure 2-2. *Change data capture (CDC)*

CDC extracts data in real-time from the source refresh. The data are sent to the staging area before being loaded into the data warehouse of the data lake. In this process, data transformation occurs in chunks. The load process places data into the target source, where it can be analyzed with the help of algorithms and BI (Business Intelligence) tools.

There are many techniques available to implement CDC depending on the nature of your implementation. They include the following:

- **Timestamp**: The Timestamp column in a table represents the time of the last change; any data changes in a row can be identified with the timestamp.

- **Version Number**: The Version Number column in a table represents the version of the last change; all data with the latest version number are considered to have changed.

- **Triggers**: Write a trigger for each table; the triggers in a table log events that happen to the table.

- **Log-based**: Databases store all changes in a transaction log to recover the committed state of the database. CDC reads the changes in the log, identifies the modification, and publishes an event.

The preferred approach is the log-based technique. In today's world, many databases offer a stream of data-change logs and expose them through an event.[5]

Data Mesh

In modern data business disruption, you need to have the right set of technologies in place to support it. Organizations are working on implementing a data lake and data warehouse strategy for datafication, which is good thinking. Nevertheless, these implementations have limitations, such as the centralization of domains and domain ownership. There are better solutions than concentrating all domains centrally; you need to have a decentralized approach. To implement decentralization, you need to adopt a data mesh concept that provides a new way to address common problems.

The data mesh helps create a decentralized data governance model where teams are responsible for the end-to-end ownership of data, from its creation to consumption. This ownership includes defining data standards, creating data lineages, and ensuring that the data are accurate, complete, and accessible.

The goal of the data mesh is to enable organizations to create a consistent and trusted data infrastructure by breaking data lakes into silos and then into smaller, more decentralized parts, as shown in Figure 2-3.[6]

---

[5] Cloud Native Architecture and Design Patterns, *Shivakumar Goniwada, Apress, 2021*

[6] Cloud Native Architecture and Design Patterns, *Shivakumar Goniwada, Apress, 2021*

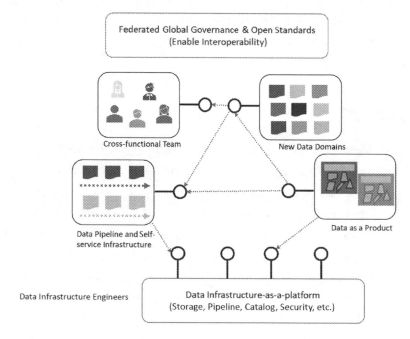

Figure 2-3. *Data mesh architecture*

To implement the data mesh, the following principles must be considered:

- Domain-oriented decentralized data ownership and architecture

- Data as a product

- Self-service infrastructure as a platform

- Federated computational governance

Machine Learning Patterns

Building a production-grade machine learning (ML) model is a discipline that takes advantage of ML methods that have been proven in the engineering discipline and applies them to day-to-day business problems.

In datafication, a scientist must take advantage of tried and proven methods to address recurring issues. The following are a few essential patterns that can help you define effective datafication.

Hashed Feature

The hashed feature is a technique used to reduce the dimensionality of input data while maintaining the degree of accuracy in the model's predictions. In this pattern, the original input features are first hashed into a smaller set of features using a hash function.

The hash function maps input data of arbitrary size to a fixed size output. The output is typically a string of characters or a sequence of bits that represent the input data in a compact and unique way. The objective of the hash function is to make it computationally infeasible to generate the same hash value from two different input data.

The hashed feature is useful when you working with high-dimensional input data because it will help you to reduce the computational and memory requirements of the model while still achieving accuracy.

The hashed feature component in ML transforms a stream of English text into a set of integer values, as shown in Table 2-1. You can then pass this hashed feature set to an ML algorithm to train a text analytics model.

Table 2-1. *English Text to Integer Values*

Comments	Sentiment
I loved this restaurant.	3
I hated this restaurant.	1
This restaurant was excellent.	3
The taste is good, but the ambience is average.	2
The restaurant is a good but too crowded place.	2

Internally, the hashing component creates a dictionary, an example of which is shown in Table 2-2.

Table 2-2. *Dictionary*

Comments	Sentiment
This restaurant	3
I loved	1
I hated	1
I love	1
Ambience	2

The hashing feature transforms categorical input into a unique string using a hash function, which maps the input to the fixed-sized output with a integer. The resulting hash value can be a positive or negative integer, depending on the input data. To convert the hash value to a positive integer, the hashed feature takes the absolute value of the hash value, which ensures the resulting index is always positive.

For example, suppose you have a categorical feature of "fruit" with "orange", "apple", and "watermelon." For this, you apply a hash function to each value to generate a hash value, which can be a positive or negative integer. After this, you can take the aboslute value of the hash value and use the modulo operator to Mal the resulting index to a fixed range of values. Suppose you want to use 100 buckets, so you take the modulo of the absolute hash value with 1000 to get an index in the range [0,99]. If the hash value is negative, taking the absolute value ensures that the resulting index is still in the range [0,99].

Using a hash function handles the large categorical input data with high cardinality. By using a hash function to map the categorical values to a small number of hash buckets, you can reduce the dimensionality of the input space and improve the efficiency of model. However, using small hash buckets can leads to collisions, where different input values are mapped to the same hash bucket, resulting loss of information. Therefore, it is important to choose a good hash function and a suitable number of buckets to minimize collision and ensure that the resulting feature vectors are accurate.

If a new restaurant opened in your area and the restaurant team launched a campaign on social media but there was no historical value existing for this restaurant, the hashing feature could still be used to make predictions. The new restaurant can be hashed into one of the existing hash buckets based on its characteristics, and the prediction for that hash bucket can be used as a proxy for the new restaurant.

Embeddings

Embeddings are a learnable data representation that maps high-cardinality data to a low-dimensional space without losing information, and the information relevant to the learning problem is preserved. These embeddings help to identify the properties of the input features related to the output label because the input feature's data representation directly affects the final model's quality. While handling structured numeric fields is straightforward, disparate data such as video, image, text, audio, and so on require training of the model. It would be best if you had a meaningful numerical value to train models; this pattern helps you to handle these disparate data types.

Usually, one-hot encoding is a common way to represent categorical input variables. Nevertheless, in disparate data the one-hot encoding of high-cardinality categorical features leads to a sparse matrix that does not work well with ML models and treats categorical variables as being independent, so we cannot capture the relationship between different variables using one-hot encoding.

Embeddings solve the problem by representing high-cardinality data densely in a lower dimension by passing the input data through an embedding layer that has trainable weights. This helps capture close relationships between high-dimensional variables in a lower-dimensional space. The weights to create the dense representation are learned as part of the optimization model.

The tradeoff of this model is the loss of information involved in moving from a high-cardinality representation to a low-dimensional representation.

The embedding design pattern can be used in text embeddings in classification problems based on text inputs, image embeddings, contextual language models, and training an autoencoder for image embedding where the feature and the label are the same.

Let us take the same restaurant example: You have a hundred thousand diverse restaurants, and you might have ten thousand users. In this example, I recommend the restaurant to the respective users based on their likings.

Input: 100,000 restaurants, 10,000 users to eat; task: recommended restaurants to users.

I have put the restaurants' names in order, with Asian restaurants to the left, African restaurants to the right, and the rest in the center, as shown in Figure 2-4.

Figure 2-4. List of restaurants

I have considered the dimensions of separating the content-specific restaurants; there are many dimensions you can consider, like vegetarian, dessert, decadent, coffee, etc.

Let us add restaurants to the x-axis and y-axis as shown in Figure 2-5, with the x-axis for Asian and African and the y-axis for Europe and Latin American restaurants.

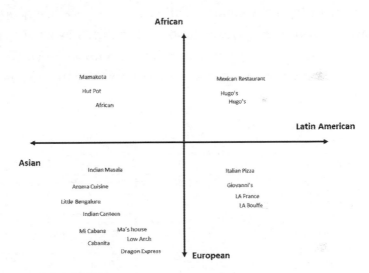

Figure 2-5. *Restaurants along x-axis and y-axis*

The similarity between restaurants is now captured by how close these points are. Here, I am representing only two dimensions. The two dimensions may need to be three to capture everything about the restaurants.

d-**Dimensional Embeddings**: Assume user interest in restaurants can be roughly explained by *d* aspects, and each restaurant becomes the *d*-dimensional point where the value in dimension *d* represents how much the restaurant fits the aspects and embeddings that can be learned from data.

Learnings Embeddings in Deep Network: No training process is needed. The embedding layer is just a hidden layer; supervised information (e.g., users went to the same restaurant twice) tailors the learned embeddings for the desired task, and the hidden units discover how to organize the items in d-dimensional space in a way to optimize the final objective best.

If you want restaurant recommendations, we want these embeddings aimed toward recommended restaurants. The matrix shown in Figure 2-6 is a classic method of filtering input. Here, I have one row for each user and one column for each restaurant. The simple arrow indicates that the user has visited the restaurant.

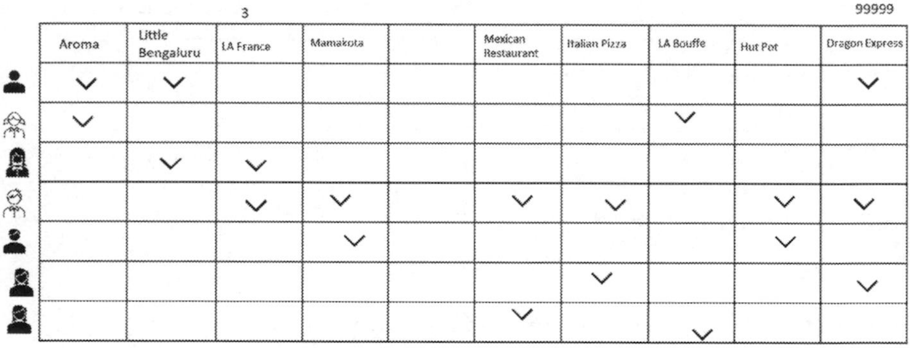

	Aroma	Little Bengaluru	LA France	Mamakota		Mexican Restaurant	Italian Pizza	LA Bouffe	Hut Pot	Dragon Express
	✓	✓								✓
	✓							✓		
		✓	✓							
			✓	✓		✓	✓		✓	✓
				✓					✓	
							✓			✓
						✓		✓		

Figure 2-6.   Restaurants matrix

Input Representation:

- For the above example, you need to build a dictionary mapping each feature to an integer from 0, ..., restaurant 1.

- Efficiently represent the sparse vector as just the restaurants at which the user dined; the representation is shown in the figure.

You can use these input data to identify the ratings based on user dining and provide recommendations.

Selecting How Many Embedding Dimensions:

- Higher-dimensional embeddings can more accurately represent the relationships between input values.

- Having more dimensions increases the chance of overfitting and leads to slower training.

The embeddings can be applied to dense social media, video, audio, text, images, and so on.

Feature Cross

The feature cross pattern combines multiple features to create new, composite features that can capture complex interactions between the original features. It is always good to increase the representation power of the model by introducing non-linear relationships between features.

The feature cross pattern can be applied to neural networks, decision trees, linear regression, and support vector machines.

There are different ways to implement the feature cross pattern depending on the algorithm and the questionnaire.

The first approach is manually creating new features by combining multiple existing features using mathematical operations. For example, a credit risk prediction task requires multiplying the applicant's income by their credit score. The disadvantages of this approach is that it is time consuming.

The second approach is automating feature engineering by using algorithms to automatically generate new features from the existing ones. For example, the autoML library can be used to generate new features based on relationships between features in each set. The disadvantage of this approach is that it is difficult to interpret a large number of features.

The third approach is neural network–based and uses neural networks to learn the optimal feature interactions directly from the data. For example, a deep neural network may include multiple layers that combine the input features in a non-linear way to create new, higher-level features.

The benefits of using feature cross patterns are as follows:

- They improve model performance by capturing complex interactions between features that may not be captured by individual features alone.

- They can use automation to generate new features from existing ones, and reduce the need for manual feature engineering.

- They help to improve the interpretability of the model by capturing meaningful interactions between features.

The drawbacks of feature cross patterns are as follows:

- Feature crosses can increase the number of features in the model, which can lead to the curse of dimensionality and overfitting.

- Feature crosses can increase the computational complexity of the model, making it harder to train and scale.

Let's consider a simple example to use feature crosses for linear regression in R.

```
library(caret)

# Load example data
data <- data.frame(
  x1 = c(1, 2, 3, 4),
  x2 = c(2, 4, 6, 8),
  y = c(3, 5, 7, 9)
)

# Define the feature cross formula
fc_formula <- y ~ x1 + x2 + x1:x2
```

```
# Train the model with feature crosses
model <- train(
  fc_formula,
  data = data,
  method = "lm",
  trControl = trainControl(method = "none")
)
# Evaluate the model
pred <- predict(model, newdata = data)
result <- sqrt(mean((pred - data$y)^2))
print(result)
```

The Result : [1] 0

In this example, I have loaded the caret package and data consisting of x1, x2, and y columns.

I have defined a feature cross using y~x1 + x2 +x1:x2. Here, x1 and x2 are separate features and x1:x2 to create a feature cross between x1 and x2.

Train the linear regression model with feature crosses using the train() method. In this method, I have specified fc_formular, data, and lm for linear regression and trControl to prevent cross-validation.

The next line in this example is to generate a prediction using predict() and evaluate the model's performance using the root mean squared error, and the result is 0.

The resulting value tells you the average difference between the predicted values and the actual values. In this example, the result is 0. It means that the predicted values are exactly the same as the actual values, which is unlikely to happen in real large data set. Here it is 0 because the data set is too small.

Multimodal Input

The multimodal input pattern involves combining and processing data from multiple sources, such as text, images, video, audio, or numerical data. This pattern helps you to extract meaningful features from each modality and integrate them to obtain a more comprehensive representation of the data, which can lead to improved accuracy in predictive modeling.

This process involves multiple steps, such as pre-processing, feature extraction, and fusion. The pre-processing involves cleaning and transforming data from each modality to ensure consistency and compatibility. The feature extraction involves extracting relevant features from each modality, and fusion involves combining extracted features from each modality to create a combined features representation.

Here's an example of how to implement the multimodel pattern with linear regression in R.

Table 2-3. *Metamodel.csv*

text	filename	y
"This is a great product!"	product1.jpg	8.5
"I was disappointed with the service."	service1.jpg	3.2
"The scenery is beautiful in this area."	scenery1.jpg	9.1
"The food was not very good."	food1.jpg	4.3
"I love this dress, it fits perfectly."	dress1.jpg	7.8

```
library(tidyverse)
library(magick)

# Load example data
data <- read_csv("MetaModel.csv")
```

```
# Load image data
images <- data$filename %>% map(image_read)
image_array <- images %>% map(image_data) %>% array()

# Preprocess text data
tokenizer <- text_tokenizer(num_words = 1000)
tokenizer %>% fit_text_tokenizer(data$text)
sequence <- texts_to_sequences(tokenizer, data$text)
sequence_matrix <- pad_sequences(sequence, maxlen = 100)

# Combine data
combined_data <- cbind(sequence_matrix, image_array)

# Train linear regression model
model <- lm(y ~ ., data = combined_data)
summary(model)

# Evaluate the model
pred <- predict(model, newdata = combined_data)
result <- sqrt(mean((pred - data$y)^2))
print(result)
```

- In this example, you need to load the tidyverse and magic packages.

- Load data 'Metamodel.csv' as shown in Table 2-3; the text contains the text data, filename contains the image filename, and Y contains target values.

- Pre-process the text data using `text_tokenizer()` to create a tokenizer and `text_to_sequences()` to transform the text into a numerical sequence matrix.

- Process the image data by loading and resizing images using `image_read` and `image_data`.

49

- Combine the pre-processed text and image data into a single matrix using cbind() and train a linear regression model using lm().

- The summary of model looks like the following:

```
Call:
lm(formula = .outcome ~ ., data = dat)

Residuals:
X1 X2 X3 X4
 0  0  0  0

Coefficients: (1 not defined because of singularities)
            Estimate Std. Error t value Pr(>|t|)
(Intercept)        1          0     Inf   <2e-16 ***
x1                 2          0     Inf   <2e-16 ***
x2                NA         NA      NA       NA
'x1:x2'            0          0     NaN      NaN
---
Signif. codes:  0 '***' 0.001 '**' 0.01 '*' 0.05 '.' 0.1 ' ' 1

Residual standard error: 0 on 1 degrees of freedom
Multiple R-squared:      1,      Adjusted R-squared:      1
F-statistic:   Inf on 2 and 1 DF,  p-value: < 2.2e-16
```

- The output shows the estimates, standard errors, t-values, and p-values for each of the co-efficients. The output indicates that the model fits the data perfectly, but the singularity problem indicates that there may be issues with multicollinearity in the data. This indicates the input data is highly correlated with each other.

- The result is [1] 4.075046: This indicates the data have better performance of a selected input data, in this example, it is a metamodel.csv.

Reframing Pattern

The reframing pattern is used to change the perspective of the problem. This pattern looks at a problem from a different perspective in order to gain new insights. This identifies and changes the frames of reference that you use to think about a particular situation. Here, frames are the mental models that are used to interpret and make sense of the globe around us. It will shape your perceptions and beliefs and can influence the way you approach the problem and decisions.

This pattern can be used to solve problems in the area of personal relationships, business, and management. The reframing pattern involves the following steps:

- Identify the frames of references for your problems; for example, a problem-focused frame to determine negative or positive aspects of the situation.

- Generate new frames of reference that could be used to think about the problem in a new way. For example, a "studying" frame of reference that focuses on the history that can be studied from a certain context.

- Evaluate the pros and cons of each frame of reference and choose the most appropriate frame for your problem.

Let's look at an example of how the reframing pattern could be used with regression analysis in R by using the number of years of experience and salary, as shown in Table 2-4.

Table 2-4. *Salary and Experience.csv*

years_of_experience	education_level	job_title	Salary
3	Bachelor's Degree	Software Engineer	30000
5	Master's Degree	Software Engineer	40000
2	Bachelor's Degree	Software Engineer	25000
7	Master's Degree	Software Engineer	50000
4	Bachelor's Degree	Data Engineer	35000
6	Master's Degree	Data Analyst	50000
3	Bachelor's Degree	Data Analyst	30000
6	Bachelor's Degree	Manager	35000
8	Master's Degree	Manager	50000
6	Bachelor's Degree	Manager	45000
10	Master's Degree	Manager	60000

In this example, I am analyzing the data set to predict the salary of an employee based on their years of experience and education level. Here, I am using the linear frame of reference that assumes a linear relationship between the predictor variables (years of experience and education level) and response variable (salary). This frame of reference assumes that the effect of each predictor variable is constant across all values of the other predictor variables.

The following R program shows how to fit a multilevel regression model to the salary and education data set shown in Table 2-4 by using the lme4 R package.

```
library(lme4)

# Load example data
data <- read_csv("salary_and_experience.csv")
```

```
# Fit a multilevel regression model with random intercepts for
job title
model <- lmer(salary ~ years_of_experience + education_level +
(1|job_title), data = data)
```

```
# View the model summary
summary(model)
```

In this code, the multilevel regression model is used to model the relationship between salary and years of experience, education level, and job title. The '(1|job_title) formula specifies that a random intercept should be included for each job title to account for the variation in salaries between various job titles.

The summary model of this data set is as follows:

```
Linear mixed model fit by REML ['lmerMod']
Formula: salary ~ years_of_experience + education_level + (1 |
job_title)
   Data: data
```

```
REML criterion at convergence: 159
```

```
Scaled residuals:
    Min       1Q   Median       3Q      Max
-1.53235 -0.53754  0.06677  0.38041  1.42926
```

```
Random effects:
 Groups    Name          Variance Std.Dev.
 job_title (Intercept)    792030   890
 Residual                 10229222 3198
Number of obs: 11, groups:  job_title, 5
```

Fixed effects:

	Estimate	Std. Error	t value
(Intercept)	19526.4	2957.8	6.602
years_of_experience	3459.0	646.5	5.351
education_levelMaster's Degree	5521.7	2833.9	1.948

Correlation of Fixed Effects:

	(Intr)	yrs_f_
yrs_f_xprnc	-0.886	
edctn_lvM'D	0.438	-0.723

Output Description:

- REML criterion at convergence: 159. This measures the goodness of the fit of the model, with lower values indicating a better fit.

- Random effects show the estimated variance components for the random intercepts of the job_title variable and the residual variation. The estimated variance of the random intercepts is 792030, which indicates that there is significant variation in salaries between different job titles. The estimated variance of the residual variation is 10229222, which indicates that there is still some unexplained variation in salaries.

- Fixed effects show the estimated coefficients for the predictor variables along with their standard errors and t-values. The intercept term estimated to be 19526.4, which represents the estimated salary for an employee with a zero years of experience and a bachelor's degree. The coefficient for years_of_experience is estimated to be 3459.0, which indicates that for each additional year of experience, the estimated salary increases by AED (United Arab Emirates Currency) 3459. The 5521.7 indicates the employee with a master's degree earns an estimated AED 5521.7.

- The correlation of fixed effects is 0.886, which indicates that employees with more years of experience tend to have a lower starting salary. The 0.438 indicates that employees with master's degree tend to have higher starting salaries. –0.723 indicates that the effect of years of experience on salary is different for employees with bachelor's versus master's degrees.

Ensemble Pattern

The ensemble is a machine learning pattern that combines several base models and aggregates their results to produce one optimal predictive model. As you are already aware, no ML models are perfect. The standard errors in ML models are irreducible, errors due to bias, and errors due to variance. The irreducible errors are the errors in the model resulting from data-set quality, the framing of the problem, or bad training examples. The other two, bias and variance, are reducible errors where you can influence or optimize the model.

This pattern is more powerful, and you can use it to improve the accuracy and robustness of predictions. However, its also requires careful tuning and can be computationally intensive.

There are several ways to implement this pattern. The choice of method depends on the specific problem and data at hand, as follows:

- **Bagging**: This involves training multiple models on different random subsets of the training data, and then aggregating their predictions to form a final prediction, because each model in the ensemble will be slightly different due to the randomness of the training data.

- **Boosting**: This involves training multiple models sequentially, with each model attempting to correct the errors of the previous model. The idea behind this is to create multiple weak models that can collectively make strong predictions.

- **Stacking**: This involves training multiple models on the same data and then using their prediction as input features to train a metamodel. The idea behind this to learn how to combine the predictions of the base models to make a more accurate final prediction.

- **Random Forest**: This involves an ensemble of decision trees, where each tree is trained on a random subset of the feature and the training data. The final prediction is then a majority vote of the predictions of each tree.

There are various techniques available in ensemble patterns to solve the problem, as we've just seen.

The ensemble pattern combines several ML models to decrease the bias and variance and improve model output. Bagging or bootstrap aggregating, boosting, and stacking are three ensemble techniques.

Let's look at an example of how to implement the ensemble pattern for linear regression in R.

The sample data set shown in Table 2-5 is used in the program.. This data contains five input variables ($a1$, $a2$, $a3$, $a4$, $a5$) and one output variable (b). The objective is to predict the value of b based on the values of $a1$, $a2$, $a3$, $a4$, and $a5$.

Table 2-5. *Ensemble Data*

a1	a2	a3	a4	a5	b
1.1	2.5	3.4	4.5	5.6	10.5
2.5	4.5	5.5	7.5	9.5	18.5
3.2	5.5	6.5	8.5	11.5	24.5
4.5	7.5	8.5	11.5	14.5	30.5
5.2	8.5	9.5	12.5	15.5	35.5

In this program, the ensemble pattern is used to combine the predictions of a linear regression model and a ridge regression model to create a more accurate prediction. The ensemble prediction is calculated by averaging the predictions of the two models on the test data. The root mean squared error (rmse) of the ensemble prediction is then calculated as a measure of prediction accuracy.

```
# Load example data
data <- read_csv("Ensemble_Data.csv")

# Split the data into training and test sets
set.seed(123)
train_indices <- sample(1:nrow(data), 0.8*nrow(data))
train_data <- data[train_indices, ]
test_data <- data[-train_indices, ]

# Fit a linear regression model to the training data
model1 <- lm(b ~ ., data = train_data)

# Fit a ridge regression model to the training data
library(glmnet)
x <- model.matrix(b ~ ., data = train_data)[,-1]
y <- train_data$b
model2 <- cv.glmnet(x, y, alpha = 0.5)
```

```
# Make predictions on the test data using the linear
regression model
pred1 <- predict(model1, newdata = test_data)

# Make predictions on the test data using the ridge
regression model
x_test <- model.matrix(b ~ ., data = test_data)[,-1]
pred2 <- predict(model2, newx = x_test)

# Combine the predictions using simple averaging
ensemble_pred <- (pred1 + pred2)/2

# Calculate the root mean squared error of the ensemble
prediction
rmse <- sqrt(mean((ensemble_pred - test_data$b)^2))
print(rmse)
```

The output is: [1] 1.409946.

The 1.409946 is the root mean squared error (rmse) between the prediction made by the ensemble model and the actual target values. This value suggests that the ensemble model is making relatively accurate predictions, with an average error of 1.409946 units in the target variable.

Chain Pattern

The chain pattern is a design pattern in software engineering that involves creating a chain of objects that can perform specific tasks. Each object in a chain has a specific role or responsibility, and the chain can handle a request by passing it along from one object to the next until it is processed completely.

In machine learning, the chain pattern can build systems that involve multiple models or processing steps. The chain pattern addresses the situation where machine learning problems can be broken into a series of steps, such as data cleansing, feature selection, feature scaling, etc.

Let's consider an example: assume that you have a list of data sets that need to be processed in a specific way depending on the column names. For a data set with a column named "A", you want to add a new column "B" that contains both "A" and "C", while for a data set with a column "B", you want to add a new column "C" that contains the "B" and the "D". In this example, I want to implement this processing in a way that is flexible and extensible, so that new processing requirements can be added in the future without affecting the existing code.

To implement this example, you can use the chain pattern, as follows:

- Define a base class called Data setHandler that defines a method handle for handling data sets. This base class will be extended by each specific data set's handler class.

- Define a specific handler class for each type of data set, such as ColumnAHandler and ColumnBhandler. Each of these classes must override the handle method.

- Connect the handlers into a chain, defining a successor relationship between them.

Here is the code to do that:

```
# Define the base Data setHandler class
Data setHandler <- R6::R6Class("Data setHandler",
                        public = list(
                          handle = function(df) {
                            stop("handle method not
                            implemented")
                          },
                          set_successor =
                          function(successor) {
                            self$successor <- successor
                          }
                        ),
```

```
                                    private = list(
                                      successor = NULL
                                    )
)

# Define specific handler classes for each data frame type
ColumnAHandler <- R6::R6Class("ColumnAHandler", inherit = Data
setHandler,

                                  public = list(
                                    handle = function(df) {
                                      if ("A" %in% names(df) & "C"
                                      %in% names(df)) {
                                        df$B <- df$A + df$C
                                        return(df)
                                      } else if (!is.
                                      null(self$successor)) {
                                        return(self$successor
                                        $handle(df))
                                      } else {
                                        return(df)
                                      }
                                    }
                                  )
)

ColumnBHandler <- R6::R6Class("ColumnBHandler", inherit = Data
setHandler,

                                  public = list(
                                    handle = function(df) {
                                      if ("B" %in% names(df) & "D"
                                      %in% names(df)) {
                                        df$C <- df$B * df$D
```

```
                        return(df)
                     } else if (!is.
                     null(self$successor)) {
                        return(self$successor
                        $handle(df))
                     } else {
                        return(df)
                     }
                  }
               )
)

# Create a chain of handlers and handle the data frames
column_a_handler <- ColumnAHandler$new()
column_b_handler <- ColumnBHandler$new()
column_a_handler$set_successor(column_b_handler)

df1 <- data.frame(A = c(1, 2, 3), C = c(4, 5, 6))
df2 <- data.frame(B = c(2, 4, 6), D = c(1, 2, 3))
dfs <- list(df1, df2)

processed_dfs <- lapply(dfs, function(df) column_a_
handler$handle(df))
print(processed_dfs)
```

Rebalancing Pattern

This pattern is used to improve the performance of machine learning models and provides various approaches for handling inherently imbalanced data sets.

Imbalanced data sets are a common scenario in classification problems, such as fraud detection, anomaly detection, and spam detection. Typically, the assumption you make is that all classes are balanced and thus result in poor predictive performance. This pattern addresses building models with data sets with few examples for a specific class.

The rebalancing pattern involves modifying the training data to rebalance the classes, making them more equal in size. This can be done in various ways, including oversampling, undersampling, and generating synthetic data.

Oversampling helps you to randomly duplicate observations from the minority class to increase its size, to rebalance the class sizes and prevent the model from being biased toward the majority class. This can be done by using various techniques, such as random sampling, stratified sampling, etc.

Undersampling is the opposite of oversampling; it involves randomly removing observations from the majority class to reduce its size. This can be done by using various techniques, such as Tomek Links, random sampling, stratified sampling, etc.

Generating synthetic data involves creating new observations for the minority class to increase its size, which helps you to balance the class size to prevent bias toward the majority class.

Once you have rebalanced training data by using these techniques, it can be used to train the model. This technique can be repeated multiple times to improve the performance of the model.

The drawback of this model is oversampling can lead to overfitting and can increase the risk of misclassifying observations, and undersampling leads to loss of valuable information from the majority class. So, you need to decide which technique to adopt based on your question.

Checkpoint Pattern

The checkpoint is a point where the model's current training state is saved. You can use this pattern to save the program state periodically to a checkpoint file in the event of failure or interruption. This can help prevent data loss and reduce the time required to recover from failures. The checkpoint pattern can be applied in a variety of contexts where the state of long-running applications can change over time, including microservices, database management systems, simulations, etc.

There are various methods available to implement the checkpoint pattern, depending on the specific question. They are periodic checkpoint, incremental checkpoint, redundant checkpoint, and versioned checkpoint.

The checkpoint helps you to minimize data loss, speed up recovery time, and improve debugging. The downsides of checkpoints are storage requirements, performance impact on your model, and complexity.

Stateless Serving Pattern

Stateless serving does not maintain any state or information about the previous request, which means that each request is independent. Stateless serving is often used in microservices and distributed architecture, where it can help to simplify the design of the system and make it easier to scale and maintain.

In the context of the machine learning model, it does not maintain the state of or information about previous requests. Once the models are trained and ready for production, you need to deploy them into the production environment and use them to make the prediction, and the model should be resilient enough to meet the demands in a real-time production environment.

The stateless serving design pattern helps to scale the infrastructure environment synchronously to handle millions of prediction requests per second in real-time. The stateless function is a function whose output is determined purely based on the inputs, stateless objects are immutable, and weights and biases are stored as a constant.

In the stateless approach, the state of the previous execution in the epoch is not referenced by the last execution. Each request sent across the model can be interpreted and does not require reference, but you need to store the checkpointing and other training details in a constant for reference. This stateless serving is similar to the behavior of stateless protocols such as HTTP(s).

The few use cases other than in microservices architecture are as follows:

- The stateless pattern can be used in model serving, which involves deploying ML models as a service that can be accessed by other applications.

- Inference as a Service helps you deploy your model as a service, allowing users to use your model to do certain predictions on new data without needing to train the model.

- Stateless pattern can be used in the model chain pattern, which involves connecting multiple models together to perform complex tasks.

Batch Serving Pattern

The batch serving pattern uses distributed data processing to execute in parallel and in many instances. It is useful when you must carry out predictions asynchronously, unlike stateless serving, where you can process one or 100+ cases for a single request.

Let us take an example of stock trading and crypto mining, where you need to use batch serving to predict large data sets. The batch serving pattern uses distributed data processing infrastructures, such as MapReduce, Apache Spark, Big Query, Apache Beam, etc., to carry out ML models on many instances asynchronously.

Continued Model Evaluation Pattern

Continued model evaluation refers to the practice of regularly evaluating the model post production. This pattern helps to handle the common problem of fit-for-purpose. Deploying into production is only part of the process; you must regularly evaluate the model's behavior and accuracy. Over time, the model can degrade, and its predictions can become increasingly unreliable. The main reasons for model degradation over time are concept drift and data drift.

Concept drift occurs when the relationship between the model inputs and the target has changed. This happens because the underlying assumptions of your model have changed. For example, your model was developed to detect credit card fraud; over time, the behavior of credit card usage also changed, like pin usage, tapping with Wi-Fi, cardless transactions, etc.

Data drift occurs when the data being fed to the model for predictions has changed compared to the data used for training. Data drift occurs for any reason: the input data schema changes, feature distribution changes over time, etc.

Continuous monitoring of the model's predictive performance over time is the way to identify the model's fitness. Whatever report comes from evaluation depends on what amount of performance deterioration is acceptable about the cost of retaining; it is a similar approach to general microservice architecture evaluation.

The key steps of this pattern are as follows:

- Collecting the new data on a regular basis, which is used to retrain models and evaluate their performance.

- Retrain a model on the new data set to update the parameters to improve the accuracy.

This pattern is useful in the area of predictive maintenance, fraud detection, and customer churn analysis, where it is important to continuously monitor and refine models to maintain their accuracy and effectiveness.

Summary

In this chapter, I illustrated datafication-relevant principles and patterns with the adoption methodology often used in datafication. The patterns related to feature engineering, model selection, model training, model evaluation, model deployment, and continued model evaluation.

Leveraging principles and patterns gives engineers a high-level structure of the datafication and provides a grouping of design decisions that have been repeated and used successfully. Using them reduces complexity by placing constraints on the design and allows us to anticipate the qualities the cloud-native system will exhibit once it is implemented.

These patterns can be applied iteratively to build and improve machine learning models over time.

CHAPTER 3

Datafication Analytics

Datafication analytics analyzes data sets to draw a conclusion and gain insights from the data. It uses statistical models and computational techniques to identify data patterns, trends, and relationships.

Every minute of every day, massive amounts of data are being generated and consumed. Whenever you post on your social media accounts, play a video, or navigate through web pages, you contribute to the data. This information forms a unique data set that can be used for marketing and business intelligence, which helps businesses make quick, effective decisions.

Datafication analytics refers not only to the volume and variety of data but also to the algorithms required to mine that data using predictive, machine learning, and prescriptive models.

This chapter covers data patterns, trends, relationship analytics, process details, metrics, and algorithms necessary for data analysis.

This chapter provides the details of the unanswered questions, like the following:

- What is analytics?

- What are the algorithms required for datafication?

- What are the details of each algorithm?

- What are the elements of datafication?

© Shivakumar R. Goniwada 2023
S. R. Goniwada, *Introduction to Datafication*,
https://doi.org/10.1007/978-1-4842-9496-3_3

Introduction to Data Analytics

Data analytics is a robust process. Every organization uses analytics to gain insights into customer behavior, market trends, patient performance, shop floor analysis, supply chain, etc. This can help organizations improve operational efficiency, identify new opportunities, and gain customer confidence.

What Is Analytics?

Seemingly unconnected data collected from various sources can be transformed into new, valuable, practical, and human-understandable knowledge.

The analysis of data to extract knowledge to inform decision making is known as data analytics. Analytics is a process that includes data collection, pre-processing, transformation, modeling, and interpretation of the final processed data. Large volumes of structured and unstructured data are collected, primarily unstructured data from social media and Internet of Things (IoT).

IDC (International Data Corporation) predicts the global data sphere will be approximately 175 zettabytes (ZB) by 2025. This number is staggering. Note that one zettabyte is roughly equal to one billion terabytes. If you compare with normal circumstances, if each terabyte were a kilometer, a zettabyte would be equivalent to 1,300 round trips to the moon. Now multiply by 175, and you will realize the big picture of the data deluge in today's business.

Big Data and Data Science

Big data refers to the enormous amounts of data collected from many sources and that is generated with increasing velocity. It consists of structured, unstructured, and semi-structured data sets that must be processed. Data processing requires analytics with volume, variety, veracity, and velocity, as follows.

- Volume is concerned with how to store large data sets.

- Variety comprises data from different sources and formats, such as audio, video, log files, and transaction history from various sources.

- Veracity is refers to accuracy, reliability, and trustworthiness of the data. it is one of the key aspects of data quality. when data is considered to have high veracity, it means that it is dependable and can be relied upon for decision-making, analysis.

- Velocity is the ability to deal with data arriving quickly in various modes of transport.

Big data is a technology that provides a computing environment for all kinds of data processing, including analytics. Various framework tools, such as Hadoop, Spark, Cassandra, Zookeeper, HBase, and MLib, are used for data processing of large-scale data.

Data science is about data analytics with advanced techniques, including data mining, statistical analysis, predictive analytics, and more, analyzing vast amounts of data and drawing a conclusion by uncovering hidden patterns and correlations, trends, and other valuable business information and knowledge.

The main goal of analytics is to help organizations make better business decisions and future predictions and analyze large numbers of transactions. For example, the major FMCG (Fast Moving Consumer Goods) e-commerce organizations will use data from social media, such as Facebook and Google, to view customer information and behavior.

The organization must remember that the success of analytics is based on the volume and veracity of data. The veracity of data is importat becahse inaccurac or unreliable data can lead to incorrect conclusions, flawed analysis, and poor decision-making. Analytics is based on the effective use of ingested data. Therefore, it is better to understand what type of data it is, how it can be used, and how it is easier to understand.

Each analyst in an organization has their preferred way of analyzing things, but your methods should be simple and up to what KPI (Key Performance Indicators) wants. The choice of analytical model is a significant decision before you start the analysis. The techniques can be as simple as a list of metrics distributed in a particular order or displayed side by side in a certain way.

Various algorithms are available for data analysis, and the use of available algorithms depends on your objective. These algorithms are broadly classified into four categories:

> **Descriptive Analytics**: This is used to summarize, visualize, and interpret data. The descriptive algorithms provide insights into past events and trends and help identify data patterns and relationships. Descriptive analytics uses a few techniques to analyze the data. They are frequency distributions, central tendency measures (such as mean, median, and mode), and actions of variability (range, variance, and standard deviation). A few examples where you can use descriptive analytics are sales data analysis, social media analytics, website traffic analysis, and customer segmentation analysis.

> **Diagnostic Analytics**: These algorithms analyze data to understand why certain events occurred. The diagnostic algorithms are used to identify the diagnosed root cause. Diagnostic analytics uses a few techniques to analyze the data. They are regression analysis, correlation analysis, hypothesis testing, root cause analysis, and more. A few examples are analyzing the root cause of quality

issues in any product, patient health diagnosis, and what factors contribute to the health of the patient. You can also use it to analyze why customers are leaving the business.

Predictive Analytics: These algorithms analyze data to predict future events and trends. Predictive analytics uses a few techniques to analyze the data. They are time series analysis, machine learning algorithms, decision trees, regression analysis, and forecasting. A few examples of predictive analytics are behavioral analysis, fraud detection in financial industries, retail inventory management, etc.

Prescriptive Analytics: These algorithms analyze data and provide future action recommendations. Prescriptive analytics uses a few techniques to analyze the data. They are simulation, decision trees, machine learning algorithms etc. the few examples where you can use prescriptive analytics are sales forecasting, supply chain optimization for any retail industry etc.

Following is the list of elements that are required for your analysis with the help of the analytical techniques:

- **Tools**: What tools will you use for research?

- **Metrics**: What type of metrics will you collect? Again it depends on the nature of the product you conduct an analysis

- **Time**: are you working on comparisons weekly, daily etc.?

- **Dashboard and Reporting**: What you want to show in the dashboard with graphs.

71

Datafication Analytical Models

Following are the analytical models most predominantly used in the datafication process.

Content-Based Analytics

Content-based analytics involves analyzing the content of documents, text, images, and videos to extract meaningful insights from unstructured data such as social media posts and reviews. This analytics helps you to identify patterns and relationships within the data.

Let's consider an example of a new product launch of consumer goods. The company wants to analyze the customer feedback on its product and what customers like and dislike about the new product. You need to harvest the data from various social media platforms. To do this, you harvest, you need using y use an algorithm such as Natural Language Processing (NLP) or any other algorithm.

By using NLP, you can analyze the sentiment of each customer review and categorize it with positive, negative and neutral sentiment on each review by using the dictionary. After categorizing all the reviews, you can get the sentiment towards your product. Based on the overall sentiment, you can decide about your product launch.

Data Mining

Data mining is the analytics process of identifying patterns and insights from the data set using statistical and machine learning (ML) techniques. You can use data mining for both structured and unstructured data.

Let's consider an example of an Auto Insurance Platform. In this platform, the team wants to analyze purchasing behavior to identify customer preferences patterns and trends by using customer demographics, product history and purchase history data sets.

To analyze these data, the team can use data mining technique such as clustering analysis or association rules. By using these algorithms team can identify relationship between data set. With association rule, team can identify which product are commonly purchased together by customers. With clustering, team required to group customers based on their purchasing behavior.

Text Analytics

The text analytics is a subset of data mining that focuses on unstructured data. the text analytics is a process of deriving meaningful information from text data and extracts meaningful information from text by using associations among entities, profiling, data reduction, clustering, self-organizing maps, predictive rules, patterns, concepts, events, etc., based on rules.

Let's consider a similar example of data mining. The company wants to launch a new television model, and before launching they want to analyze customer opinion about the model and its features. To conduct an analysis, you need to harvest social media data related to this television model and analyze it by using sentiment analysis and entity recognition techniques.

Using sentiment analysis, you can capture the emotions in the text to identify positive or negative sentiments toward the new television.

By using entity recognition, you can identify and categorize named entities like influential people, organizations, and locations of any discussion of this model. This helps you to identify the list of influencers to rope them in to promoting this television model.

Sentiment Analytics

Extensive unstructured or semi-structured data are available on social media, such as transaction data, blogs, and news websites. These data are used to understand what people say and feel about products, brands, and so on. Using natural language processing (NLP) or by parsing phrases or sentences, semantic analysis can detect sentiment and reveal changes in sentiment to predict possible scenarios. The goal is to automatically recognize and categorize opinions expressed in the text to determine overall feelings. The critical aspects of sentiment analysis are polarity classification, emotion detection, characteristics, and multilinguality.

Graded Sentiment Analysis

Polarity refers to the overall sentiment conveyed by a particular text, phrase, or word. This polarity can be expressed as a numerical rating known as a sentiment score.

The polarity categories are as follows:

- Very Positive
- Positive
- Neutral
- Negative
- Very Negative

Each category is assigned grades 5, 4, 3, 2, and 1, respectively.

Emotion Detection Analysis

Emotion detection sentiment analysis allows you to go beyond polarity to detect emotions such as happiness, frustration, and anger. Many emotion detection systems use lexicons or complex ML models.

Aspect-based analysis

When analyzing the sentiment of texts, you want to know which aspects or features people mention in a positive, negative, or neutral way.

Multilingual Analysis

Multilingual analysis is complex because it involves considerable pre-processing and resources. The language in texts can be detected by using the language classifier API in the cloud.[1]

Audio Analytics

Audio analytics is a process that allows you to analyze audio data by using various ML algorithms. It involves analyzing audio data such as voice, audio streams, and more to extract relevant information based on the question.

Audio analytics refers to extracting meaning and information from audio signals for analysis. Audio processing removes acoustic features relevant to the task, followed by decision-making schemes involving detection, classification, and knowledge fusion. You can use various approaches, including the Gaussian mixture–universal background model (GMM–UBM), support vector machines (SVM), and the deep learning neural network.

The following are the categorizations of audio analytics:

- **Extracting Nonverbal Cues from Human Speech**: Analyzing a human voice to extract information beyond speech recognition, including speaker identification.

- **Audio Understanding**: The aim is to analyze and extract insight from audio signals, such as detecting audio events and recognizing audio backgrounds and anomalies.

- **Audio Search**: This helps to search the audio and is essential for navigating large amounts of raw audio data and metadata.

---

[1] Social Media Analytics Strategy, *Alex Goncalves, Apress, 207 [[or 2017??}}*

Video Analytics

Video analytics is a group of unique models that automatically analyze video content. The algorithms process videos to examine the content. There are three common types of video analytics, as follows:

- **Fixed Algorithm Analytics**: This uses an algorithm designed to perform a particular task and looks for a specific behavior; for example, crossing a line at a traffic signal, loitering, etc.

- **Action Recognition**: This involves identifying and classifying human actions within a video stream. You can use this type of analytics in sports analysis applications, etc.

- **Facial Recognition Systems**: This involves identifying and verifying the identify of individuals within a video. ML algorithms help to match faces against the available database for border security, access control, etc.

- **Crowd Movement**: This involves analyzing video data to extract information about the movement, behavior, and characteristics of a crowd in any procession or at VIP events.

Comparison in Analytics

Analysis comparison in analytics is done to compare various metrics and key performance indicators (KPIs) to identify differences and similarities between them.

Let's take an example. Assume your product A has 1,000 likes, and you are going to compare it with another product B that you already had in mind, and product B likes are approximately 10,000, so you think 1,000 likes is low, or the product is distributed in limited places.

If you have nothing in mind for comparison, you establish that what we see is the first detail you will use. Moreover, the details are just one part of the process, and they need to be connected with other processes to draw a conclusion. The details provide insights into the specific data points, but it is necessary to analyze the underlying factors that contribute to those data points to form a conclusion.

Usually, in datafication you will compare via percentages. Nevertheless, it is essential to highlight that a percentage alone can often have no actual value and can give you wrong information.

For example, the 'A' national television channel and 'B' national channel grew 20%, let's say the initial audience size of 'A' national television channel was 1 million viewers, and the initial audience size of the 'B' national television channel was 10,000 viewers. if both channel grew by 20%, the absolute growth would be, For channel 'A' - Initial Audience: 1 million viewers, 20% growth, 1 million *20% = 200,000 viewers, the final audience are 1.2 million. For channel 'B', initial audience are 10,000 viewers, 20% growth, 10,000*20%=2000 viewers, the final audience of channel 'B' are 12,000 viewers. The conclusion is, the growth percentage is same for both the channels, the absolute growth in terms of viewers is significantly different.

Let us consider another example. The A English daily paper grew by 10 percent, and, in comparison, another B English daily paper grew by 5 percent. This comparison makes sense, but you need a base number to measure 10 percent and 5 percent, respectively.

Comparison is the core of data analysis. If you are stuck at any point in time, go back to the root cause analysis of the issue and take corrective actions by adjusting the values to make sure that your analysis provides a meaning result, as I explained in the percentage example.

Datafication Metrics

As I mentioned earlier, metrics are the primary basis of analysis. When basic information is gathered in big data, it is transformed into metrics by using various models; without metrics, there is no analytics output/input.

It is important to choose metrics that are relevant to the business or organization's objectives so as to provide meaningful insights. These metrics should be clearly defined and measurable to ensure that they can be tracked and evaluated over time, and you must regularly review and update the metrics to ensure that they remain relevant and useful in making data-driven decisions.

The metrics can be straightforward, such as "change in viewers on YouTube channels over time" and "the speed of turbines in industry during peak hours over the years."

Metrics are vital for analysis, and you cannot generate metrics at one go; they will mature over time, and many metrics can be grouped into one for better analysis.

Metrics help you to see what you have in terms of data and what you have in terms of output by using quantitative measurement of the performance of the algorithms. These will help you to quantify the accuracy, precision, recall, and other important parameters.

There are two kinds of metrics you might come across: default metrics and custom metrics.

The default metrics are pre-defined metrics that are commonly used to measure the algorithms. These metrics help you to access the accuracy, precision, and other important parameters.

Some tools provide an option to create your metrics. With this help, custom metrics can be created.

The following are a few examples of metrics:

- Follower count over time

- Location, gender, age

- Viewers of national television

- Type of post over time

- Sponsored content

- Questions and answers

The metrics you collect are only helpful if you choose the right graph or dashboard to present them. The following are the key elements when deciding if a particular graph type fits well with a particular metric:

- Quick reading

- Easy presentation

- Grouping

- Period of analysis

- The number of metrics grouped

- Color theme

For a more detailed analysis of the type of data sources, refer to Chapter 4, "Analytics in Social Media," in *Social Media Analytics Strategy* by Alex Goncalves and published by O'Reilly.

Datafication Analysis

Before getting into the details of algorithms, the reader must understand that social media analysis differs from traditional in-house data analysis. Social media or IoT analysis involves the competitive information highly available on social media and from sensors, while conventional analysis deals with enterprise data.

Traditional business intelligence, a.k.a. algorithms, provides analysis and reporting of structured data to describe past trends; as technology progresses, only some algorithms are used to predict future behavior, but not with accuracy. As the cloud emerges, data capacity also increases, and the processing power of machines is increased. Data scientists have integrated methods from statistics, mathematical equations, pattern recognition, deep learning, uncertainty modeling, and visualization to gain insight and predict behaviors based on the data.

The datafication is brought into data warehousing, and visualization and algorithm models are used to provide forward predictive capabilities, which are based on the behavioral activities of users in social media and on IoT sensors.

Datafication requires technology to process a sizeable extensive data set efficiently and involves machine learning, signal processing, simulation, natural language processing, time series analytics, and so forth.

Data Sources

Before starting analytics, it is good to understand the data source and how it works. This will give you a fair idea of how to apply the best algorithms. Providing data sources and their details is important to ensure transparency, reproducibility, and accountability in datafication. This helps you to ensure that the data being used is accurate, reliable, and relevant to the metrics being studied. See the following:

- **Offline Data**: This are data that have been generated and stored outside of an online environment. These data can be collected in a variety of ways, such as through paper surveys, interviews, and observations. They are used to supplement online data and provide additional context that helps organizations to make datafication decisions.

- **Online Data**: This is data generated with an internet connection; for example, websites, e-commerce apps, social media applications, etc.

For datafication, you need to combine both offline and online data sources to make data-driven decisions. By combining rich qualitative insights of offline data and with quantitative online data, you can gain a more complete understanding of customer behavior and preferences.

There are various data sources available in social media, IoT, health-care systems, retail platforms and so on. The health-care system and retail platform data are directly sourced from hospitals, digital devices, and controlled users from e-commerce platforms, but there are diverse sources and data from various social media platforms and IoT devices, a few of which are listed here:

- **Post and Comments**: Any contents that users post on social media platforms, such as text, images, videos, and links. You can use posts and comments to analyze behavior, preferences, and opinions.

- **Likes and Shares**: The actions that users take on social media platforms, such liking, sharing, or commenting on posts. You can use these data to analyze the popularity of the particular content.

- **Followers and Fans**: Number of people who follow or like social media accounts. You can use these data to analyze the popularity of the account.

- **Hashtag**: Keywords or phrases used to categorize social media content. By tracking hashtags you can analyze the relevance of any product in social media.

The data sources of IoT devices are as follows:

- **Industrial Sensors**: Used in industry to monitor machinery and track inventory. You can use these data to analyze the health of the machine and to optimize inventory.

- **CCTV**: Used in tracking people or vehicle movement, you can use these data to identify behavior of the customer or vehicle movements.

- **Health-care Devices**: The data from remote
  monitoring systems, medical devices, etc. You can use
  these data to analyze the patient's health progress.

Human activities on social media include creating a profile, publishing content, commenting on content (claps, likes, hearts, dislikes, etc.), sharing content, restricted sharing within groups, sending group chats, private chats, online shopping, and so on.

Although many features are standard across the social network when performing an analysis, there are specific challenges related to each social platform in terms of availability, mode of availability, and interpretation of data. For example, Twitter's likes differ from those of Facebook.

It is essential to understand the data in terms of facts and estimates. Take the example of paid media. There are estimated metrics, including views, impressions, and audience reach (which provides the number of times an audience has potentially seen a particular piece of content or advertising). Sometimes these metrics may only be accurate if someone has seen the range if there is a bidirectional channel from the audience. You might have seen the news channels and they usually put in big and bold letters, "Our channel viewers are 50% more than others." You cannot question and trust these numbers because they are invisible.

Another point of view on data is whether the data are public or private. Without going into a deep discussion on categorizing public and private, the public data are the data anyone can see when navigating the platform. The confidential data are data only the owner of which can view. The publicly available data can be compared, but private data cannot be reached; it can be estimated.

For a more detailed look at the analysis type of data sources, refer to Chapter 4, "Analytics in Social Media," in *Social Media Analytics Strategy* by Alex Goncalves and published by O'Reilly.

Data Gathering

The data from social media platforms or IoT gateways can be gathered in three ways

- Via API (application programming interface)

- Streaming

- Web crawling or scraping

The easy way to collect the data is via API and streaming, as crawling is complicated due to network restrictions. These data are called raw data and are loaded into the landing zone. Apply a few rules to cleanse and store it in the data warehouse for analytical purposes.

Introduction to Algorithms

The key characterstics of algorithms are:

1. Well-defined inputs and outputs

2. Precise and unambiguous instructions

3. Finite number of steps or instructions

4. Effective and efficient to solve the problem

The algorithms can be viewed as a union of unsupervised learning methods, more commonly known as data mining, and supervised learning methods deeply rooted in mathematical theory, specifically statistics, optimization, etc. The third part is reinforcement learning, where goal performance is earned but has yet to be explicitly recognized.

Machine learning (ML) is a process that involves the use of a computer to make decisions based on data. The approaches of ML are broadly grouped into the following three categories:

- Supervised Machine Learning

- Unsupervised Machine Learning

- Reinforcement Machine Learning

Figure 3-1 shows the breakdown of different machine learning models.

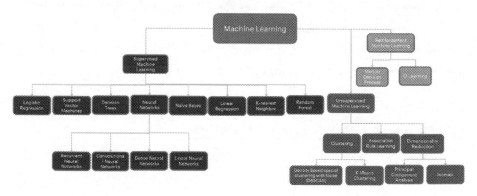

Figure 3-1. *Machine learning algorithms*

Supervised Machine Learning

Supervised machine learning is an algorithm that can produce patterns and hypotheses by using externally supplied instances to predict the fate of future cases and aims to categorize data from prior information. It is trained using labeled data sets to classify data or predict outcomes accurately. As input data are fed into the model, it adjusts its weights until the model has been fitted appropriately, which occurs as part of the cross-validation process.

A supervised ML algorithm aims to find a mapping function to map the input variable (x) with the output variable (y).

Supervised learning uses the training data to teach models to yield the desired output, as shown in Figure 3-2. These training data include inputs and correct outputs, which allow the model to learn over time. The algorithm measures its accuracy through the loss function, adjusting until the error has been sufficiently minimized. The functioning of supervised learning is shown in Figure 3-2.

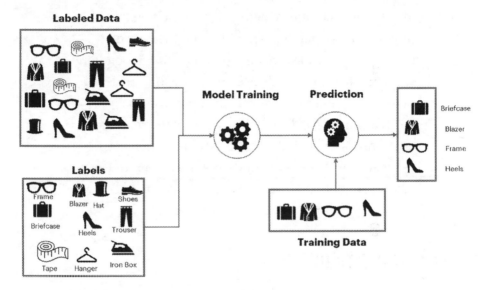

Figure 3-2. *Supervised machine learning*

Note   Icons are from Microsoft Icons Library.

The first step is to train the model for various icons. If the given shape is an eye frame and all sides are equal, it will be labeled as a frame.

Supervised learning can be separated into two types of problems:

- **Classification**: This involves predicting a categorical or discrete output variable based on the input variables. The classifier algorithms learn the model that can accurately predict the targeted variables based on the input features. The performance of the classification algorithm is typically evaluated using metrics parameters, such as accuracy, precision, recall, and F1-score.

- **Regression**: This is used to understand the relationship between dependent and independent variables. Regression techniques mainly differ based on the number of independent variables and the type of relationship between the independent and dependent variables. It is commonly used in forecasting and determining the cause-and-effect relationships between variables. The performance of the regression algorithm is evaluated using metrics parameters such as mean squared error (MSE), root mean squared error (RMSE), and R-squared. The standard algorithms are linear regression and logistic regression.[2]

Linear Regression

The Linear Regression algorithm is used to identify the relationship between a dependent variable and one or more independent variables and is typically leveraged to make predictions about future outcomes. It is known as simple linear regression if there is only one dependent and one independent variable. Suppose two or more independent variables are thought to be related to the dependent variable; this is known as multiple linear regression. Each type of linear regression seeks to plot a line of best fit, as shown in Figure 3-3. For social media and IoT types of prediction, linear regression is one of the best fits.

---

[2] https://www.coursehero.com/ *Data Analysis*

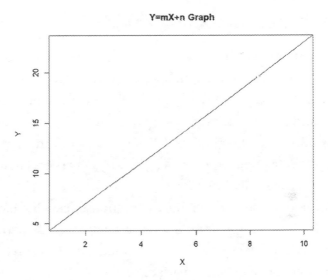

Figure 3-3. *Linear regression model*

The regression line with Y=mx+n will be generated with these few lines of R script:

```
#Create a data frame with X and Y Values
data <- data.frame(X = 1:10, Y = 2*1:10 + 3)
# Plot Y vs X with line
plot(data$X, data$Y, type="l", main="Y=mX+n Graph", xlab="X",
ylab="Y")
#add the Y=mx+n line to the plot
abline(3, 2)
```

In Figure 3-3, the center line is the best-fit line, and it represents the linear relationship between the independent and dependent variables in the data. The best-fit line is also called the regression line or the line of best fit.

$$Y=mX+n$$

The relationship between *X* and *Y* can be defined by *m* and *n*. Given enough data sets, your objective will be to find a suitable *m* and *n*.

Support Vector Machines (SVM)

The SVM is used for both classification and regression. This algorithm finds a linear or non-linear decision boundary between two data variables. SVM aims to find a hyperplane that maximally separates the data into classes. The hyperplane is chosen so that the margin between the two types is maximized. The margin is the distance between the hyperplane and the nearest data points from each class. The data points that are close to the hyperplane are called vectors.

The SVM can be used for both linear and non-linear classification tasks. In linear, the decision boundary is a straight line or hyperplane that separates the data into two classes. In non-linear, the decision boundary is a non-linear function of the input variables.

The SVM can be used in the classification of email spam or ham (unwanted or legitimate) classification of credit risk as high or low.

The SVM can be used in regression analysis to predict continuous output variables based on one or more input variables. An example of SVM regressions is the prediction of plot or apartment prices.

Here is a simple example of making predictions based on the SVM:

```
# Load the required packages containing the SVM()
library(e1071)
# Generate some example data
set.seed(123)
# example data with 40 observations with two predictor
variables 'x' and binary response variables 'y' to -1 for the
first 20 observations and 1 for the last 20 observations
x <- matrix(rnorm(40*2), ncol=2)
y <- c(rep(-1,20), rep(1,20))
# Train the SVM model, with the 'x' and 'y' data as inputs.
C- classification indicates binary classification problem, and
linear indicates linear decision boundary.
```

```
model <- svm(x, y, type="C-classification", kernel="linear")
```

```
# Make predictions for new data
New data <- matrix(rnorm(20*2), ncol=2)
predictions <- predict(model, new data)
```

```
# Print the predictions for the new data based on the trained
data. The predicted class labels are binary values.
print(predictions)
```

The predictions are:

```
1  2  3  4  5  6  7  8  9 10 11 12 13 14 15 16 17 18 19 20
 1 -1  1 -1  1 -1 -1  1  1 -1 -1 -1  1  1 -1  1 -1 -1  1  1
Levels: -1 1
```

Decision Trees

The decision tree covers classification and regression but is preferred in classification problems. The analysis uses a decision tree to represent decisions and decision making visually and explicitly. As the names indicate, it uses a tree-like model, where internal nodes represent the features of a data set, branches represent the decision rules, and each leaf represents the outcome.

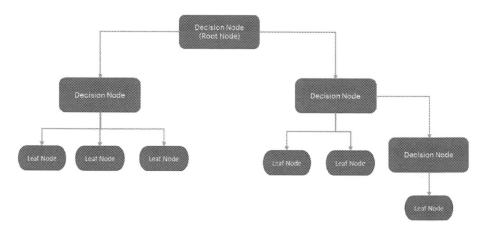

Figure 3-4. *Decision Ttee machine learning model*

As shown in Figure 3-4, a decision tree has two nodes: the decision node and the leaf node. Decision nodes are used to make decisions and have multiple branches, leaf nodes are the output of those decisions, and there are no other nodes after the leaf.

Let us consider the pandemic and the chances of surviving COVID-19-infected patients. Figure 3-5 uses the condition of the person who has pre-existing diseases and is in the ICU.

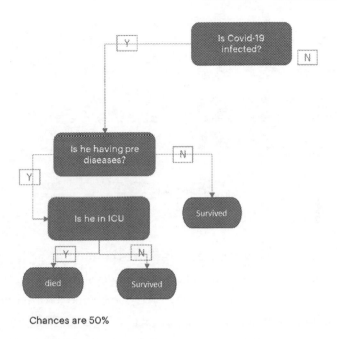

Chances are 50%

Figure 3-5. *Decision tree model*

Decision trees are generally trained using a heuristic. This easy-to-create algorithm gives a nonoptimal but closest optimal decision tree, primarily using a divide-and-conquer strategy to train an algorithm. At each node, all the possible conditions are evaluated and scored.

You can get more decision and leaf nodes in the actual data set with various conditions.

Neural Networks

A neural network is an ML algorithm that processes data as the human brain does. It is a type of deep learning that uses interconnected nodes or neurons in a layered structure that resembles the human brain. It creates an adaptive system that computers use to learn from their mistakes and improve continuously.

91

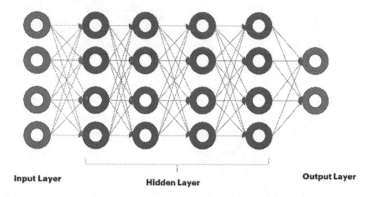

Input Layer

Hidden Layer

Output Layer

Figure 3-6. *Neural network*

The neural network comprises node layers containing an input layer, one or more hidden layers, and an output layer, as shown in Figure 3-6. Each artificial neuron in a neural network is connected to other neurons through weighted connections. The output of each neuron is computed as a weighted sum of the inputs, where the weight represents the strength of the relationships between the neurons. The result of the neuron is then passed through an activation function, which introduces non-linearity into the model and allows the neural network to learn complex patterns in the data.

The input layer is an entry point for the information into the neural network. The input nodes process the data, analyze and categorize it, and pass it into the next layer.

The hidden layers take the input from the input layers. Neural networks can have a large number of nodes in hidden layers. Each hidden layer analyzes the output, processes it further, and passes it to the next layer.

The output layer produces the final data processing results in the neural network. It can have single or multiple nodes depending on the information you process. For example, the output layer will have one node if you have a binary classification problem with 0 or 1. If you have a multiclassification problem, the output layer might have more than one node.[3]

---

[3] https://www.coursehero.com/ *Data Analysis*

Naïve Bayes Algorithm

Naïve Bayes is a classification approach that adopts the principle of class conditional independence from the Bayes theorem.

The Bayes theorem is named after Thomas Bayes and describes the probability that an event will occur based on prior knowledge and conditions of circumstances. For example, retail sales on a Black Friday or Deepavali in India are known as an event. Bayes' theorem allows the sales of items for stores to be associated more accurately than random prediction. This means that the independence between events implies that the presence or absence of one feature or event does not impact the probability of another event occurring and occurrene of one event does not provide any information or influence the likelihood of the other event happenning in the probability of an outcome.

There are three types of Naïve Bayes classifiers: multinomial naïve Bayes, Bernoulli Naïve Bayes, and Gaussian naïve Bayes.

Bayes' Theorem

As stated, Bayes' theorem finds the probability of an event's occurring given the likelihood of another event that has already happened.

$$equalP\left(A|B\right) = \frac{P\left(B|A\right) * P\left(A\right)}{P\left(B\right)}$$

where A and B are events and $P(B) \neq 0$. Here, I am finding the probability of A, given that event B is true and termed evidence. See the following:

- **P(A|B)**: The probability of hypothesis A with the data or event of B, which is called the posterior probability

- **P(B|A)**: The probability of event B given that hypothesis A was true

- **P(A)**: The probability of hypothesis *A*'s being actual, which is called the prior probability of *A*

- **P(B)**: The probability of the event's occurring regardless of the hypothesis[4]

K-Nearest Neighbor (KNN) Algorithm

KNN is a nonparametric algorithm that classifies data points based on their proximity and association with other available data. The algorithms assume that similar data points can be found near each other and are used for classification and regression problems.

The KNN algorithm assumes that similar things exist nearby, or, in other words, similar things are near each other, as shown in Figure 3-7. The KNN algorithm stores all the available data and classifies a new data point based on the similarity. When new data appear, they can be classified into a category.

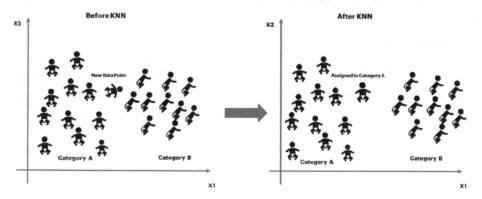

Figure 3-7. *KNN model*

As an example, consider the preceding data sets containing two baby positions. With the help of the KNN algorithm, you can quickly identify the category of a particular baby part.

---

[4] *Coursehero.com, Data Analysis coursehero*

The KNN involves the following steps:

- The KNN algorithm selects the number of K neighbors.

- It calculates the Euclidean distance of K neighbors.

- It takes K's nearest neighbors as per the Euclidean calculation.

- It counts the number of data points in each category.

- It assigns a new category to which the number of neighbors is maximum.

Random Forest

The random forest algorithm is used for both classification and regression problems. As the name indicates, it is a collection of uncorrelated decision trees merged to reduce variance and create more accurate data predictions by combining many decision trees.

Let us consider an example of an opinion poll prediction of an election, as shown in Figure 3-8. The decision tree starts with "Who will win this election?" From there, you can ask questions based on the basic living facilities, benefits, freebies, etc. These questions make up decision nodes in the tree, acting as the means to split the data. Each question helps an individual voter arrive at a decision, denoted as a leaf node, observing the criteria that follow a particular party. The classification and regression tree (CART) algorithm trains the decision tree.

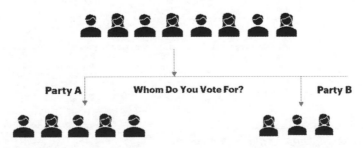

Figure 3-8. *Random forests*

Imagine that the figure data set consists of a mix of elderly, young, male, female, and LGBT people who desire to vote for any party based on their needs and likes. A few people in my group voted for Party A, and a few voted for Party B.

The random forest consists of decision trees (here, I am showing only one, but you can include more decision trees based on your example) and operates as an ensemble. Each tree in the random forest spits out a class prediction, and the class with the most votes becomes our model's prediction.[5]

Unsupervised Machine Learning

Unsupervised machine learning uses an algorithm to analyze unlabeled cluster data sets. These algorithms discover hidden patterns or data groups without the need for human intervention and training data sets. Unsupervised learning cannot be directly applied to a regression or classification problem. Unlike supervised learning, you have the input data but do not have corresponding output data.

Let us consider the same example used in supervised learning to identify a product type. The task of unsupervised learning is to identify the products' images on their own, as the respective algorithm is never trained upon the given data set, which means it needs to learn about the features of the items. It performs this task by clustering the image data set into groups according to image similarities.

---

[5] *Coursehero.com, Data Analysis coursehero*

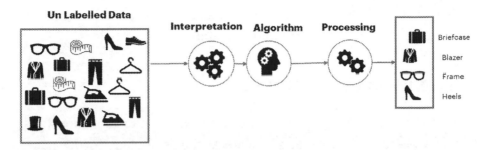

Figure 3-9. *Unsupervised machine learning*

In Figure 3-9, I have taken unlabeled input data, which means it is not categorized, and the corresponding output is also not given. These unlabeled data are fed to the model to train it. First and foremost, it interprets the raw data to find the hidden patterns in the data and then applies suitable algorithms.

Unsupervised learning models are used in clustering, association, and dimensionality reduction.

Clustering

The clustering algorithm groups unlabeled data based on similarities or differences. It is used to process raw, unclassified data objects into groups represented by structures or patterns in the information. Clustering consists of techniques to group the data points into different clusters. Composed of similar data points, the objects with possible similarities remain in a group that has fewer or no similarities with another group. It does this by finding similar patterns in the unlabeled data sets, such as shape, color, size, behavior, etc.. It divides them according to the presence and absence of those similar patterns.

Let us consider the retail shop example. You find items or products grouped, such as shampoos, T-shirts, frozen food, vegetables, etc., so you can find them easily. The clustering technique also works similarly.

The clustering technique is used primarily in social network analysis, anomaly detection, image segmentation, statistical data analysis, etc.

97

There are different categories of clustering, such as exclusive, overlapping, hierarchical, fuzzy, and probabilistic.

Association Rule Learning

Association rule learning is a rule-based method for finding relationships between variables in each data set. It checks the dependency of one data item on another and maps accordingly. This method uses different rules to discover the intriguing relationships between variables and data sets.

This method is frequently used in market-based analysis, allowing companies to better understand the relationships between different products. A well-known example is e-commerce's "Customers who bought this item also bought this other item" message for customers.

The association rules work with similar concepts of "if" and "else" statements; for example, if A, else B.

The association rules use algorithms such as Apriori, Eclat, and FP-Growth.

The Apriori algorithm is the most widely used. It uses frequent data sets to generate rules and is designed to work on databases containing transaction data; for example, sales data. This algorithm uses the breadth-first search and has a tree to calculate the products efficiently.[6]

Dimensionality Reduction

This technique is used when the data set is too high. This algorithm reduces the number of data inputs to a manageable size while preserving the data set's integrity. This algorithm aims to remove irrelevant, redundant, or noisy features from the data set. There are two main types of dimensionality reduction: feature selection and feature extraction.

---

[6] *Coursehero.com, Data Analysis coursehero*

The feature selection technique objectively removes irrelevant or redundant features from the data set. This is done by evaluating the importance or relevance of the feature to the target variable.

The feature extraction technique aims to create new features or variables that capture the most critical information in the original data set.

This algorithm is commonly used with IoT data, such as signal processing and manufacturing hubs.

Reinforcement Machine Learning

Reinforcement is training machine learning models to make a sequence of decisions. It is about taking suitable action to maximize reward in a particular situation. Use effective machines and various software to find the best possible behavior or path for a specific action. The agent receives positive feedback for each excellent effort, and for each lousy step, the agent gets negative feedback. In this model, the agent learns automatically based on the input without any labeled data.

Let us consider a few examples to help you to understand better:

- A chess player makes a move, and the choice of action is informed by thinking and anticipating possible moves by an opponent.

- A tic-tac-toe game learns the moves based on the best possible routes.

Beyond the agent and the environment, there are four reinforcement learning elements: a policy, a reward signal, a value function, and, optionally, a model.

- A policy defines the learning agent's behavior at a given time. It is a mapping from perceived states of the environment to actions to be taken in those states.

- A reward signal defines the goal in a reinforcement learning problem. At each time step, the environment sends a single number of rewards to the reinforcement learning agent, and the agent's sole objective is to maximize the rewards.

- The value function specifies what is good in the long run. The total reward depends on the value of the state.

- The model mimics the behavior of the environment. Given the state and action, the model might predict the resultant next state and the next reward.

The Markov decision process and Q-learning algorithms are part of reinforcement learning.

Summary

I illustrated the processes and various analytics types in this chapter and introduced machine learning models.

You need an ML model such as text analysis, machine learning, deep learning, or natural language processing to analyze the various analytics mentioned in this chapter. Businesses can analyze the data and gain insights, making significantly better and faster decisions.

The analytics helps you uncover hidden patterns, correlations, market trends, behaviors, preferences, hypotheses about the behavior, and so on that allow organizations to decide based on data points.

Machine learning models, such as supervised, unsupervised, and reinforced models, help one to use suitable models to predict knowledge.

CHAPTER 4

Datafication Data-Sharing Pipeline

In today's digital age, data has become a valuable commodity that helps with decision making and innovation across industries. As a result, collecting, storing, and analyzing data have become increasingly important. However, the value of the data lies not only in its collection but also in its sharing.

The data-sharing pipeline is a vital component of the datafication process. It involves collecting and preparing data for sharing, identifying potential sources and targets of the data, and implementing mechanisms to transfer the data securely. Practical data-sharing pipelines require consideration of ethical, legal, regulatory, and technical aspects to ensure that the data is shared securely.

Data sharing is not a new concept; individuals, organizations, and governments have been sharing information since way back, even before the advent of digital technologies. In earlier days, data sharing was primarily done through manual means, such as paper surveys, letters, and other types of documents. However, the evolution of computers, digital technology, and storage has dramatically changed the way data is shared, making it faster and more efficient. As a result, data sharing has become a crucial part of any business, including health care, machinery, finance, retail, and so on.

© Shivakumar R. Goniwada 2023
S. R. Goniwada, *Introduction to Datafication*,
https://doi.org/10.1007/978-1-4842-9496-3_4

Before the rise of big data and streaming techniques, data movement was typically a linear process that involved moving data to the data warehouses or operational data sources through a batch process with a sequential approach. However, with big data and streaming techniques, data movement has become more complex and sophisticated. Modern data movement techniques involve real-time data ingestion and processing, and data can be moved between various data sources.

Datafication has upended the traditional approach to data transfer, and as a result modern data transfer systems must be prepared for scale, speed, complexity, security, reliability, and flexibility.

While I was writing this book, more than 70 percent of global data and analytics decision makers were expanding their capability to use social media data, according to the Forrester report "Business Technographics Data and Analytics Survey."

Gartner predicts that by 2023, organizations that promote data sharing will outperform their peers on most business-value metrics. However, Gartner predicts that by 2023, less than 6 percent of data-sharing programs will correctly identify trusted data and locate trusted data sources.

This chapter provides insight into modern data-sharing techniques for datafication, looks at why data sharing is critical to datafication, and examines how delivery models could increase value for data providers and consumers.

Introduction to Data-Sharing Pipelines

Present-day enterprises have to contend with exponentially increasing volumes of batch, streaming, and synchronous APIs (Applicaiton Program Interface). These comprise various structured, unstructured, and semi-structured data types and originate from the expanding number of disparate sources located on-premises, in the cloud, and at the edge. At the same time, datafication users demand faster and easier access to reliable, trusted, up-to-date data to make accurate decisions.

Effective data sharing requires a thorough understanding of these considerations and the implementation of appropriate policies, procedures, and tools to ensure that data sharing is done responsibly and securely. The data pipeline refers to collecting, preparing, and sharing data. Data privacy and security refer to protecting sensitive data from unauthorized access, and data availability ensures that data is accessible and reliable. Data lineage is the ability to trace the origin of data and understand its transformation over time, while metadata provides context and additional information about the data being shared.

Until recently, only a few large organizations were using datafication in the decision-making process. Real-time data sharing is critical to efforts to obtain valuable insight into the decision-making process; however, many organizations still use old, outdated methodologies to discover access data from platforms and devices. For datafication, you need to use internal and external data to generate business insights. For both types, you require a well-defined methodology and governance.

Steps in Data Sharing

Several steps can be followed when sharing data; these steps may vary depending on the type of analysis and context.

The following are the high-level steps:

- **Determine the purpose and usage of data**: This step helps to determine the most appropriate method for sharing the data.

- **Identify the data source and obtain necessary approvals**: These steps help identify the data source. For example, if your usage is a medical diagnosis, the data source is hospitals or clinics. If your usage is product reviews, the data source could be social media.

- **Prepare the data for sharing**: This step involves cleansing and organizing data, adding metadata, and ensuring that data are in a format that can be easily accessed and understood.

- **Create a pipeline architecture and identify suitable platforms**: Create a data pipeline for extracting, loading, and transforming data to the target platform.

- **Manage metadata**: Information about data, such as its source, format, and quality, that helps others to understand and use the data.

- **Data license type**: The term and conditions under which the data can be used, such as whether it is open access or restricted to specific users.

- **Document the data-sharing process**: Create a necessary document of the data-sharing process, including source, approvals, or permissions obtained.

- **Data management**: The process of organizing, storing, and curating the data, including backups, business continuity, and ensuring that the data are easily accessible.

Data-Sharing Process

As mentioned in the preceding steps, data may come from many sources, including web analytics; Internet of Things (IoT) for digital twin (It is a virtual representation or digital replica of a real-world physical object); social media platforms, such as Twitter, Meta, Google; medical devices, and more, depending on context and usage. Datafication involves various mechanisms for sharing data, including file transfer protocols, APIs, cloud storage, streaming, and virtualization. The choice of the data-sharing agent depends on the specific needs and requirements.

The following are the most commonly used sharing mechanisms in the datafication process:

- **Application Programming Interface (API)**:
  APIs enable different applications and systems to communicate and exchange data with each other. APIs provide a standard interface for accessing and manipulating data.

- **Extract, Load, and Transfer (ETL)**: Extracts data from the provider's database, transforms it so it is suitable for consumption, and loads it into the database

- **Change Data Capture (CDC)**: Transfers data from one source to another by using a change data mechanism through streaming

- **Cloud Storage**: The provider stores a copy of the data and exposes the data to the consumers.

- **Streaming**: Streams the data from various sources by using streaming software

- **Batch**: Bulk batch files from various sources using a scheduler and secured file transfer

- **Virtualization**: Virtualization is a logical data layer that integrates data within and outside of an enterprise across disparate systems and delivers business users in real-time.

Modern data platforms can serve as the control center for sharing data among critical business applications without copying the data locally. In this process, the data provider, such as the IoT Edge gateway or social media platforms, will grant access to the data to query with the defined sharing mechanism. Once you query the data, you need to apply the pipeline for processing the data.

Data-Sharing Decisions

The need to manage data movements efficiently is a primary driver for the data pipeline since datafication requires obtaining data from various diverse sources. If appropriately managed, moving data can be manageable for IT resources and capabilities.

The data pipeline is central to the emerging area of datafication. It describes processes related to the movement and consolidation of data between data providers and data stores. The pipeline consolidates data into consistent forms. It is the most basic functionality that every enterprise needs in order to move data.

The data pipeline is critical to the datafication process because it focuses on transforming and integrating data from the source system to consolidated data hubs or lakes, and from those hubs or lakes to the data analysis engine, where it can be delivered to consumers for their dashboard.

Datafication seeks to integrate various data types whether in a structured or an unstructured format.

The following is the decision process you need to follow before designing:

1. You need to have a business process in order to create a data-sharing pipeline; the business goals are as follows:

    •   Lower the cost and complexity of managing the data-sharing pipeline.

    •   Provide data securely with regulatory compliance

    •   Data must be available in real-time or in batch mode.

    •   It must support diverse integration mechanisms and diversified data.

    •   It should support business intelligence, analytics, etc.

2. Without proper inputs, you may not be able to design an effective pipeline; the following are the few inputs that help you to develop the pipeline:

    - Data semantics

    - Data source type

    - Regulatory compliance, standards, and security

    - Data processes, data requirements

3. It is mandatory to define domain ownership for each set of data, as follows:

    - Data and their relevant components should be owned, maintained, and developed by the team closest to them.

    - Determine the owner who has the authority to decide how data are collected, used, and shared.

    - Clearly define the roles and responsibilities of those who handle and manage data, including data owners, custodians, and users.

    - Create policies and standards to govern data collection, data storage, and data retention, including guidelines to access the data.

    - Define and communicate data ownership policies across all relevant stakeholders.

    - Always adhere to the data privacy policy and collect and use data according to local regulations.

    - Establish a good governance structure to oversee data ownership use. It should not be on the centralized level of the organization.

Data-Sharing Styles

Traditionally, enterprises' data approaches require more time to manually prepare data with batch-only ingestion and no standardized data sets, and also to orchestrate data movement between sources and targets. Finding the correct patterns and making data available based on business needs is the right balance to create a proper data pipeline.

Unidirectional, Asynchronous Push Integration Style

This style refers to the data flowing in one direction from a source to a target. This means that the data is transferred from data sources such as medical devices, social media, IoT sensors, etc., but any changes made to the target destination system are not reflected in the original. This type of integration is often used when the source system is read-only; it is utilized extensively in datafication, as shown in Figure 4-1.

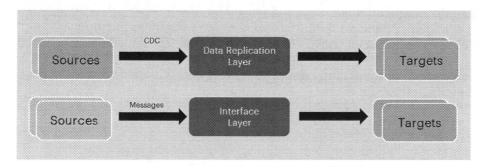

Figure 4-1. *Unidirectional, asynchronous integration style*

Asynchronous push refers to data flow that is not in real-time. There will be a time lag between data generation and data processing. This type of pattern will be used when the systems involved have different processing speeds or when the data being transferred are not critical to real-time operations, as shown in the preceding figure.

Real-Time and Event-based Integration Style

Real-time event-based data integration involves integrating and synchronizing data from source systems in real-time. This type of integration can be accomplished using event-driven architecture, messaging systems, and event-driven software. Event-based integration focuses on integrating data from multiple applications within an event-driven ecosystem. Integration is triggered and supported by event-driven data movement with the help of publish/subscribe, queuing, and streaming paradigms by using an event broker, as shown in Figure 4-2.

Event-based integration allows one data provider to take a subset of what it does and share it with other data management software. You also need to consider that how data providers access their data may differ. One data source may use REST, another may use a custom API, while another might provide nightly batch files on a managed file transfer. Additionally, you may need to aggregate unrelated data from many sources in datafication, as shown in Figure 4-2.

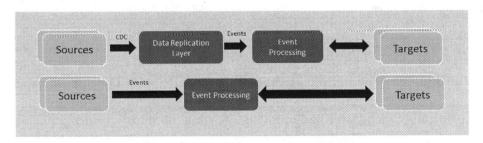

Figure 4-2. *Real-time event-based integration style*

Use cases of real-time event-based data integrations are as follows:

- Streaming data from IoT sensors and social media and processing it in real-time for analytics or machine learning

- Integrating data from various sources to create a single source of truth

- Data integration from various government entities to create domain-specific data lakes

Bidirectional, Synchronous, API-led Integration Style

In some places, you need a bidirectional integration to receive and update data from sources. Bidirectional integration is a process of integrating data from multiple sources in both directions, allowing for the flow of information between different data sources, as shown in Figure 4-3. This helps improve the accuracy of the systems in an enterprise.

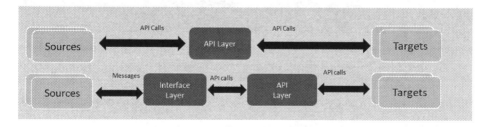

Figure 4-3. *Bidirectional, synchronous, API-led integration style*

API-led integration, as shown in the figure, is a common approach for synchronous real-time integration. This approach makes data available through APIs that different systems can access. This can be useful when there is a dependency between systems and data needs to be consistently up to date.

This style is functional when integrating IoT sensor networks and boilers in the chemical industry. The IoT sensor can send data on temperature and humidity to the boiler engine in real-time, allowing the boiler to adjust operation accordingly. The boiler can then send data back to the IoT sensor network, allowing the sensor to adapt its reading based on current conditions.

Mediated Data Exchange with an Event-Driven Approach

Mediated data exchange involves sharing information between two or more parties through an intermediary or middleman. This intermediary can be a computer program, an API, or an event-driven. It transfers data by providing a standard format, protocol, or interface, as shown in Figure 4-4.

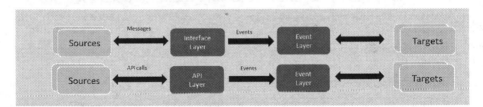

Figure 4-4. *Mediated data exchange with event-driven style*

As a classic example, this style could be used in the innovative home system; in an intelligent home system, multiple devices and sensors generate data that needs to be processed and analyzed in real-time to optimize energy efficiency and improve the overall user experience. In this case, the mediated data exchange system is an intermediary between different devices and sensors in the intelligent home system. The system can be designed to trigger events based on changes in data, such as temperature, occupancy, or home appliances or lighting. These events can then be used to activate other devices or trigger automated actions.

Designing a Data-Sharing Pipeline

Managing data movement efficiently, accurately, and securely is a primary driver for the data integration pipeline. Datafication requires managing hundreds of thousands of databases and data sources within and across organizations and has become the most important responsibility of an enterprise.

The data pipeline should be specified differently for an enterprise solution and an individual solution because of the scale and complexity of the data management.

The enterprises establishes the standards, governance framework, and security framework, and each solution or projects within an enterprises must follows the standards, this process significantly reduces the amount of time and cost spent because assessment and stakeholder discussions have been performed in advance. Developing point-to-point solutions between data sources and applications can result in thousands to millions of jobs or interfaces and it can quickly overwhelm the capabilities of even the most effective and efficient IT organizations. Figure 4-5 illustrates the end-to-end data integration pipeline design.

A data pipeline consists of multiple steps, such as data transformation, where raw data are cleansed, filtered, masked, aggregated, and standardized into an analysis-ready form that matches the target source and requirements. A data pipeline can be set up to replicate data from social media, IoT sensors, medical devices, and so forth to a data warehouse or a data lake for analytics.

The components of the data pipeline are as follows:

- **Data Source**: A data source is the point of entry for data from all sources in the pipeline. Most pipelines have transactional processing APIs, IoT sensors, batches, data collectors, etc.

- **Data Extractors**: The data extractors are protocols that extract data from different sources through API, MQTT, batch, streaming, etc.

- **Data Integration**: Data integration is a movement of data from data sources to the target source, along with the transformation performed. The ETL is the most used approach for data integration.

- **Data Storage**: Storage resides in different stages of the process where data are preserved.

- **Data Processing**: This includes activities and steps for ingesting data from sources, storing it, transforming it, and loading it into the next stage of the whole process.

- **Data Monitoring**: Monitoring ensures that the pipeline and all its stages are working as expected.

- **Data Quality**: Data quality is about data measurement, with factors such as accuracy, completeness, consistency, and reliability.

- **Data Lineage**: The data lineage covers the complete data flow lifecycle. It is the process of recording and visualizing data as it flows from the data source to the target. It provides all transformations the data underwent along the way—how the data are transformed, changed, and why it was changed, etc.

- **Data Security**: Data security is about securing your data for the data file and storing it with security techniques such as tokenization, encryption, and masking.

- **Metadata**: Metadata is data descriptions about data and enriches data with information that makes it easier to find, use, and manage.

Figure 4-5 shows an end-to-end data-sharing pipeline.

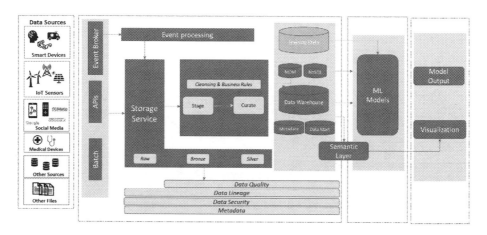

Figure 4-5. *End-to-end data-sharing pipeline*

The pipeline works, and now you can build it in your organization for datafication. The following steps will help you to build one.

In the first step, *data source* in the data sharing pipeline refers to any system, application, or device that generates or stores data that needs to be shared with another system. Data sources can include social media, IoT sensors, smart devices, medical devices, databases, and files, as shown in Figure 4-5.

In the second step, data integration types are used in the data-sharing pipeline. This type of integration mechanism, as shown in Figure 4-5, depends on the type of use cases and requirements of the data-sharing pipeline. Some options are as follows:

- The ETL technique involves extracting data from source systems and transforming it into a format that is compatible with the target system; it is typically used for batch.

- Data virtualization is an integration technique that involves creating a virtual layer on top of data sources, allowing data to be accessed and queried as if it were a single data source.

- API data integration is the grating of data from different systems where APIs are created. API integration typically involves creating API connections, establishing API contracts, etc.

- Streaming data integration typically uses eventing tools such as Kafka, and cloud-native services of hyperscale to integrate data in near real-time.

In the third step, the data are stored with a secure, scalable, and reliable way to store and manage the data by applying various transformation rules. There are different levels of data refinement in the data pipeline to manage and derive insights from the data sets more effectively. In the data storage, there four different levels of data refinement and processing, as follows:

- **Raw Data**: Raw data is the original data that is collected from various sources by using multiple data integration techniques, and this data has not undergone any pre-processing or refinement. The raw data is in the most granular forms, such as sensor data, social media feeds, etc.

- **Bronze Data**: Bronze data is the first level of data refinement in the data pipeline. It involves cleaning and transforming the raw data into a structured format ready for future processing. Bronze data is typically stored in the data lake.

- **Silver Data**: Silver data is the second level of data refinement in the data pipeline. It involves further processing the bronze data to generate insights and analytics. The silver data is typically stored in a data warehouse.

- **Gold Data**: Gold data is the final level of data refinement in the data pipeline. It involves applying advanced analytics and ML techniques to the silver data to generate predictive models. This data is used for BI (Business Intelligence) and the decision-making process (not mentioned in the diagram, this data is part of warehouse).

While processing the data from raw to silver to gold, you need to apply various transformation rules. These transformations use various techniques and tools to ensure the data is accurate, consistent, and reliable. During the transformation, you need to carry out several operations, which are data mapping, data aggregation, data filtering, data normalization, and data enrichment.

In the fourth step, you need to check the data quality. You need to implement data quality checks and validations, such as data profiling, data cleansing, and data enrichment with the help of data quality standards and policies, to ensure that the data is accurate, complete, and consistent.

In the fifth step, data lineage is to be established using various tools and technologies like integration platforms, data lineage tracking tools, and metadata management tools. The data lineage helps you to trace the origin, transformation, and movement of data throughout its lifecycle. It involves how data is created, modified, stored, and used across different systems. The data lineage is a very important component of the data-sharing pipeline because it provides a complete understanding of the data, including its history, provenance, and quality.

In the sixth step, to ensure the application is properly secured in the data-pipeline process, *security* refers to the measures and practices used to protect the data from unauthorized access and use, sensitivity, confidentiality, and compliance. The key capabilities of data security are access control, data encryption, data masking, and data tokenization.

In the seventh step, metadata is established. Metadata describes other data. It provides information about the data, like structure, format, content, and context, and enables users to search, discover, and use it more efficiently.

In the eighth step, the training data is used to train data analysis algorithms. This is a subset of the data used to teach models to recognize patterns, classify data, or make predictions. The training data is useful in your analysis to adjust model parameters to minimize the difference between the predicted output and the actual output. Data is typically labeled, which means that each data set is associated with a label.

In the ninth step, data marts will be created based on the requirements from the stakeholders. These are created by extracting a subset from a data source, such as a data warehouse or a data lake. The design of the data marts depends on specific stakeholder needs in order to provide relevant information; for example, a sales data mart can be created to provide sales teams with access to data on sales opportunities, sales revenues, etc.

In the tenth step, a semantic layer is formed. It is an abstraction layer that resides between visualization and data sources. It provides simplified, business-friendly view of underlying data, making it easier for users to access and analyze the data without needing more technical information.

In the eleventh step, the ML models use data to predict and provide trends and insights from the data based on the question. These predictions and insights can be used by business units to make informed decisions.

Types of Data Pipeline

The data pipeline is a series of processes that are used to collect, transform, and analyze data. There are several types of data pipeline that are used to handle different types of data processing. The choice of data pipeline depends on the specific use cases and the nature of the data sources. Batch processing may be more appropriate for historical analysis, and real-time or streaming processing may be more appropriate for applications that require real-time prediction. You can use a hybrid pipeline that can produce both batch and real-time processing. The following are the few data pipelines:

- Batch Integration with Managed File Transfer (MFT)

- Extract, Transform, and Load (ETL)

- Extract, Load, and Transform (ELT)

- Streaming (Event Processing)

- Lambda Architecture

- Kappa Architecture

- Data Mesh

- Data Federation and Virtualization

Batch Processing

Batch processing is the method of data processing used to complete high-volume, repetitive data jobs periodically. This type of data pipeline helps you to transfer a large volume of data in batches at various frequencies. The batch pipeline extracts data from the source using a data collector or SQL operation and transfers it to the source at regular intervals using scheduled jobs. The batch processing pipelines extract and ingest into a more extensive system in data lakes for later processing.

Batch processing is used in various scenarios, from simple data transformations to more complete ETL pipelines where the load and computation take significant time.

The batch processing architecture has data storage, batch processing, analytical data storage, analysis and reporting, and logical orchestration components.

Extract, Transform, and Load Data Pipeline (ETL)

ETL is a process of data sharing that involves extracting data from various data sources and transforming it into a format that can be loaded into a target source.

Central to all areas of data integration techniques, the ETL is the basic process to consider in the data-sharing pipeline, whether executed physically or virtually, in batch or streaming or API. These are the essential steps for moving data around and between various data sources.

As shown in Figure 4-6, the extraction process in the ETL selects the required data and extracts it from its source. The extracted data are then staged in a physical data store, either in object storage or a physical database.

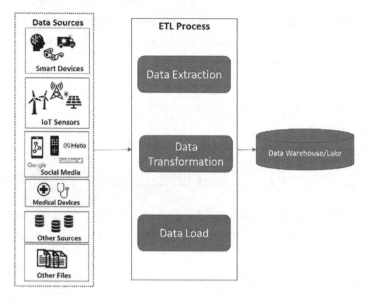

Figure 4-6. *ETL process*

The transform process makes the selected data compatible with the structure of the target data store. The transformation includes format changes, structure changes, semantic conversion, and reordering.

The load physically stores or presents the result of the transformation in the target source. Depending on the transformation performed, the transformation type, and the target system's purpose and usage, the data may need further processing to be integrated with other data.

Extract, Load, and Transform Data Pipeline (ELT)

The ELT is similar to the ETL, and the difference between them is the sequence arrangement. In the ETL, transformation is carried out on the data to match the target data source. In the ELT, the data are sent directly to the data warehouse or data lake and then transformed into the desired format. The ELT is functional when determining the data type and how to transform it. In this architecture, the transformation is done partially or entirely at once.

Streaming and Event Processing

Some call it stream processing, others call it event streaming, complex event streaming, or CQRS event sourcing, but all are the same process with different names.

Before understanding event streaming, it is good to have a clear point of view on what events are.

Anything that occurs in systems is an event; for example, customer requests, health record updates, sensor signals, etc. These events occur constantly everywhere irrespective of the type of system. Knowing about an event allows you to react to it quickly, and the more quickly you can get information about events, the more effectively your business can react to it.

The events and messages are two separate things: the event is an occurrence, and the message is the carrier of the event details that relay the occurrence.

Processing an Event

Events are recorded in an event log, as shown in Figure 4-7, and then processed by one or more services. In the event-driven architecture, events represent significant occurences or changes in teh system or domain. These events are logged and stored in a structured manner, often using a message broker or event streaming platform. The events are organized by topics or categories based on their nature or purpose, allowing interested parties to subscribe and listen to specific topics they are intersted.

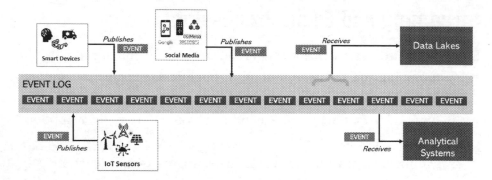

Figure 4-7. *Streaming process*

The streaming data pipeline ingests the data as it is produced and updates it constantly in response to every event that takes place. In streaming, the data are ingested from the source into eventing technology, such as Kafka, Solace, Confluent, and cloud-native streaming, as the data are created in the source and then streamed from event technology to the target source.

For data streaming, you can choose either mediator topology or broker topology, depending on the event you process. The rule of thumb to choose the topology for your use case is as follows:

- The **broker topology** can be considered for a single event or chain of events and requires one task as the result.

- The **mediator topology** can be considered for use in multiple tasks in a response that requires orchestration of each task.

The mediator topology is similar to orchestration in an SOA (Service Oriented Architecture) ESB (Enterprise Service Bus) platform; you use mediator topology when orchestrating multiple steps within an event through a central mediator. This topology is better suited for complex situations where multiple steps are required to complete the process. Thus, event processing requires coordination or orchestration.

In the broker topology, message flows are distributed across the event processor components in a rope fashion through lightweight message brokers. It does not have a central component that controls the orchestration across processes as provided by the mediator topology. The broker topology consists of a dumb broker and an intelligent processor with dump and pipe patterns.[1]

Change Data Capture (CDC)

The CDC is a replication solution that captures database changes as they happen and delivers them to the target database using a streaming technique. The CDC starts by taking a snapshot of the data on the publisher database and copying that to the subscriber database, as shown in Figure 4-8. The changes on the publisher side are sent to the subscriber in near real-time using streaming.

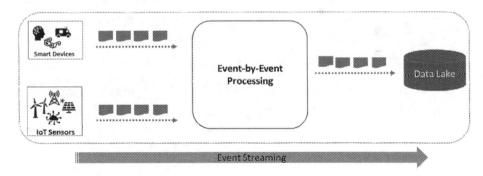

Figure 4-8. *Change data capture*

The subscriber applies the data in the same order as the publisher so that transactional consistency is guaranteed for publication with the same subscription.

---

[1] Cloud Native Architecture and Design Patterns, *Shivakumar Goniwada, Apress, 2021*

There are many techniques available to implement CDC depending on the nature of your implementation; they are timestamp, version number, triggers, or log-based approach, as follows:

- **Timestamp**: Timestamp column in a table. The timestamp column represents the time of the last change. Any data changes in a row can be identified with the timestamp.

- **Version Number**: Version number column in a table. The version column represents the version of the last change, and all data with the latest version number are considered to have changed.

- **Triggers**: Write a trigger for each table. The triggers in a table log events that happen to the table.

- **Log-based**: Databases store all changes in a transactional log to recover the committed state of the database. The CDC reads the changes in the log, identifies the modification, and publishes an event.

The most preferred approach is the log-based technique. Today's databases offer a stream of data change logs and expose these through an event.[2]

Lambda Data Pipeline Architecture

The Lambda architecture is a reference architecture for data processing and can be designed to handle data volume by using batch and streaming. Nathan Marz first introduced this architecture to help when data processing for large volumes with batch and near real-time stream processing, as

---

[2] Cloud Native Architecture and Design Patterns, *Shivakumar Goniwada, Apress, 2021*

shown in Figure 4-9. In data processing use cases, latency, fault tolerance, consistency, and scalability are mandatory non-functional requirements; the lambda architecture tries to address these important requirements.

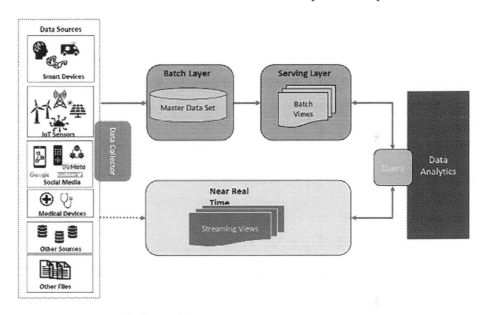

Figure 4-9. *Lambda architecture*

The main components of a lambda architecture are the data source, batch layer, serving layer, near real-time layer and query layer.

- The batch layer saves all data coming into the system as batch views; the data are treated as immutable and append-only to ensure a trusted historical record of all incoming data. The objective is to maintain all accuracy by being able to process all available data when generating views.

- The server layer incrementally indexes the batch views to make a query by data analytics or visualization; you can customize the indexes depending on the use cases. The primary purpose of the serving layer is speed.

- The near real-time layer processes data streams in real-time and handles the data that are not already delivered to the batch layer due to latency.[3]

The components within the lambda architecture complement each other, and the batch and serving layers continue to index the incoming batch file data. The near real-time layer complements the batch layer by indexing in real time all the new and delayed batch indexes. Once the batch indexing is complete, the newly indexed data will be available for analytics and visualization.

Kappa Data Pipeline Architecture

The Kappa architecture is a reference architecture for data processing for analytics and is used for processing streaming data. Jay Kreps is a CEO of Confluent introduced the reference architecture. This architecture aims to process real-time and batch processing for analytics with a single technology stack, as shown in Figure 4-10. It is based on a streaming immutable architecture in which data are stored in a database. The stream engine reads the data, transforms it into an analytical format, and stores it in an analytical database for query and data analytics and visualization.

---

[3] Cloud Native Architecture and Design Patterns, *Shivakumar Goniwada, Apress, 2021*

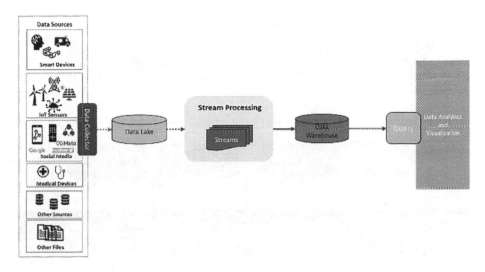

Figure 4-10. *Kappa architecture*

The Kappa architecture provides real-time analytics based on data availability. This helps the business team to reduce the decision time. It also supports historical analytics by reading the data stored in the data lake in the batch process.[4]

Data as a Service (DaaS)

The common challenges any organization faces while exchanging data usually result from fragmented data environments and data sources needing common standards. These systems' data entities and attributes often share different syntax and semantics. DaaS can enhance the implementation of data services with data integration platforms. The true potential of DaaS is based on how the data connect platform is designed with various integration patterns, as shown in Figure 4-11.

---

[4] Cloud Native Architecture and Design Patterns, *Shivakumar Goniwada, Apress, 2021*

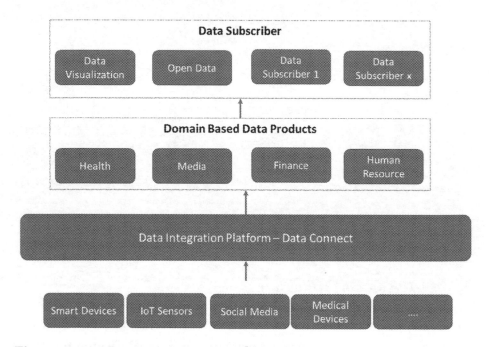

Figure 4-11. *Data as a service architecture*

The services in the DaaS (Data as a Service) must fulfill the needs of the subscriber with a few common characteristics:

- The service must provide clear value to the subscriber.

- The services must ensure the quality of the data.

- The service must enable strong data governance.

- The service must ensure the security and privacy of the data.

- The services must ensure a correct data catalog.

Data Lineage

The data lineage provides the details of the data lifecycle. Lineage software tools aim to show the complete data flow from data providers to the data store. Here, I am not going to explain it in detail, but rather provide an overview of the use of data lineage in data sharing.

Data lineage is the process of understanding, recording, and visualizing data as it flows from data sources to consumption, including data transformation steps. It provides the audit trail of the data at a granular level to help data engineers debug any data errors.

Modern data lineage tools give clear visibility into the source and journey of data. Let us take an example of medical records and how data lineage helps you trace any errors back to the root cause, as shown in the Figure 4-12.

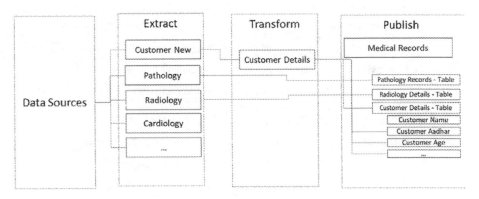

Figure 4-12. *Data lineage process*

The data lineage creates a data mapping framework by collecting and managing metadata from each step and storing it in a metadata repository that can be used for lineage analysis. For each process applied to data in its journey, the metadata will be updated in the repository.

The data lineage can help you with the following:

- Discover, track, and correct data in the data sharing process

- Lower the operation cost and ease of maintenance

- Ease of managing regulatory compliance

- Increases the data quality and governance

- Helps to understand the relationship between different data assets

- Helps with data auditing

Data Quality

Data quality involves applying quality management techniques to proactively monitor, measure, and improve the fitness of data. The data quality ensures that data can be quickly processed and analyzed.

The key functionalities of data quality are as follows:

- **Data Profiling**: This is an activity and process to analyze and assess the quality of data by using multiple statistical techniques that investigate the structure and content of data, e.g., inconsistency formats, nulls, min/max values, etc.

  Data profiling consists of multiple analyses that work together to evaluate your data, as follows:

  - **Structure Analysis**: Structure analysis helps to validate the data for consistency and format by using pattern matching and other techniques. For example, if you have a data set of Aaadhar numbers in India or SSNs in the USA, pattern matching helps you find valid sets of formats within the data set. Pattern matching can be used to identify whether a field is a number or a string.

- **Content Analysis**: Content analysis helps you identify the data quality of each respective field, looking for null values or incorrect values, etc.

- **Relationship Analysis**: Relationship analysis helps define the relationship between data sets. This process uses the metadata to determine key relationships between data and specific fields, particularly for data overlaps.

There are many techniques used today that fall under structure, content, and relationship analysis. The techniques are as follows:

- **Column Profiling**: The column technique scans through a table and counts the number of times each value is available in each column. Column analysis summarizes the result for each column, such as statistics and inferences about the characteristics of data. It identifies the invalid and incomplete data values, data class, and format expression for the values in the data.

- **Cross-Column Profiling**: Cross-column profiling assesses the relationship between tables with foreign keys and primary keys.

- **Baseline Profiling**: Baseline profiling techniques compare the multiple analysis results for the same data source for quality checks.

- **Data Cleansing**: This is an activity creating data quality cleansing rules to convert the data into a usable form based on the data profiling outcomes.

- **Dimension Monitoring**: Data quality dimensions such as completeness, uniqueness, consistency, validity, and conformity of data

Data Integration Governance

Data integration governance consists of procedures, policies, and standards to manage data-sharing integration for various sources in datafication. The objective of data integration governance is to ensure that data integration is properly managed to ensure that data are accurate, consistent, and secure from source to target.

The guidelines for data integration are as follows:

- **Data Mapping**: Create an integration blueprint for all the sources with data lineage, data protocols, and the data integration process.

- **Data Validation**: Create data validation rules that meet the standards.

- **Data Quality**: Create a guideline for quality rules that is accurate and complete with data lineage, cleansing, and monitoring in place.

- **Data Security**: Create data security guidelines for role-based access (RBAC), regulation and compliance, breaches, etc.

- **Data Housekeeping**: Create an archiving and backup policy with hot and cold storage.

You can find more details in the Data Governance Chapter 9.

Summary

Data-sharing integration connects and integrates data from various sources to make it useful in datafication. This involves multiple steps, such as data integration, which uses various data integration styles and reference frameworks. The main goal of the integration process in datafication is to enable an enterprise to make better use of data decision making.

In this chapter, you have learned various integration patterns, techniques, and reference frameworks and learned what additional processes are required in the integration.

CHAPTER 5

Data Analysis

Data analysis provides valuable insight based on the objective and questions you choose. To address these goals and questions, you need a framework consisting of steps, patterns, relationships, and trends in the data.

Donald Knuth mentioned the relationship between art and science in a whitepaper, "Computer Programming as an Art." He wrote, "Science is knowledge which we understand so well that we can teach it to a computer; and if we don't fully understand something, it is an art to deal with it. Since the notion of an algorithm or a computer program provides us with a practice test for the depth of our knowledge about any given subject, going from an art to a science means learning how to automate something."[1]

For data analysis, there are various models and tools available, but using them in the right and procedural way is essential. This chapter provides guidelines, procedures, and steps to address data analysis problems.

---

[1] *http://www.paulgraham.com/knuth.html*

© Shivakumar R. Goniwada 2023
S. R. Goniwada, *Introduction to Datafication*,
https://doi.org/10.1007/978-1-4842-9496-3_5

Introduction to Data Analysis

Data analysis is analyzing and interpreting data by using statistical and computational methods to extract meaningful insights, trends, and patterns.

Data analysis is complex, but it's not like there aren't people doing data analysis regularly. Data analysis is like the workmanship of a sculpture. The data without science is of no use in analysis, and you need to apply sound science to the data to give it life. Data analysis is valid only when the correct data are used to answer the question.

John Turkey is a leading mathematician and statistician once said, "The combination of some data and an aching desire for an answer does not ensure that a reasonable answer can be extracted from a given body of data." He also said, "An approximate answer to the right question is worth a great deal more than a precise answer to the wrong question." The meaning is you might have terabyte of data, but only need a small quantity of data to answer the fundamental data analysis question.

Through data integration, you may get an answer to struture of data process by combining data from different sources to create an unified and consistenct view of data. Still, understanding and matching this correlation and structure to specific data analysis questions is complex.

Imagine you were to ask a sculptor for a statue. There are many things the sculpture needs to consider, like how beautiful the statue should be, sketch a statue's design, what type of stone, how long it should take, how to make an armature, how to add texture, curing, etc. Mel Bouchner is a conceptual artistic and according to him According to Bochner's sculptor theory, "Any material is replaceable without changing the intention. When an object loses uniqueness, identity is denied an equivalence with presence. Any individual piece exists only as an example itself. Without the object, there is no idea. Without the idea, there is no object."

At some point, the sculptor must use creativity to combine all the required tools to make a beautiful statue.

The meaning is that you need to have good art and science to have good data analysis. Like sculpture tools, data analysis also has tools, like linear regression, non-linear regression, decision trees, deep learning, etc., as Chapter 3, "Datafication Analytics," mentions.

Data Analysis Steps

As I mentioned earlier, data analysis can be more straightforward, and it can require many iterations with a series of steps to arrive at the result for the question. As defined by Jeff Leek, he is a chief data scientist, there are seven steps, as shown in Figure 5-1, involved in data analysis. The learning on each step will be a decision for the next step.

Figure 5-1. *The core steps for data analysis*

Prepare a Question

I am giving so much importance to the question because the type and quality of the question are directly related to the data sets you choose and statistics you generate, so they are essential in preparing a question.

This step is critical in data analysis. In this step, you identify the objective and outcome you are looking for. Based on the type of question, you will identify data sets and models. The type of question is directly related to your output. There are six types of questions published by Roger D. Peng and Jeff Leek in *Science* magazine, as follows:

- Descriptive

- Exploratory

- Inferential

- Predictive

- Causal

- Mechanistic

Figure 5-2 shows a flow that will help you to identify the type of question you want to frame.[2]

---

[2] *https://www.iecodesign.com/blog/2015/4/8/clarifying-the-data-question*

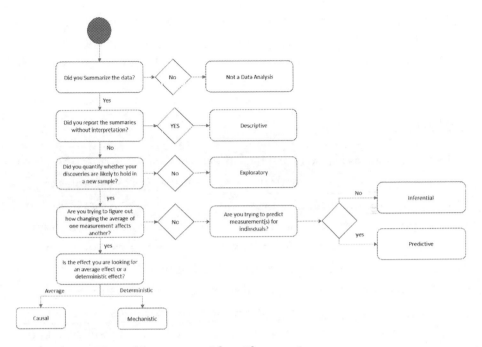

Figure 5-2. *Flow diagram to identify questions*

Descriptive

Descriptive questions are the type of questions that describe the characteristics of data sets. They seek to describe a particular aspect of the data set and are used in exploratory data analysis. Here are some examples of descriptive questions:

1. How many hotels are in New York?

2. How many vegetarian restaurants are in Dubai?

3. Number of Covid patients in July 2020? What is the distribution age of the sample in France?

In this type of question, you don't need any interpretation from the data set, but directly provide information from the attributes of the data sets. These questions help in the research field to identify specific patterns and trends.

Exploratory

Exploratory analysis is also called hypothesis analysis. The question seeks to identify any trends, patterns, and relationships in variables of general thoughts. For example, if you had a broad view of a diet linked to specific sports activity, you could explore this hypothesis by examining the relationship between sports activity and dietary factors. You might find in exploratory analysis that a particular diet increases the fitness of the persons in certain sports. As another example, what is the correlation between age and income? You could explore this hypothesis by examining the relationship between age and income. These questions are designed to be open-ended and not to achieve defined findings from the data.

Inferential

An inferential question is prepared, on restatement of a hypothesis analysis and estimation of exploratory analysis, to check or answer with a different set of data. For example, what dietary sports are present in a sample? Is there any specific diet that improves the performance of the sportsperson? By analyzing this different set of data from all over Europe in another way, you will be able to infer what is true, on average, for the frame population.

Predictive

The predictive question predicts future usage based on current and past data. Considering the same example of sports, the question could be, how many probable sportspersons undergo dietary changes in a specific country or region or across geographies?

Causal

Causal questions are questions about the relationship between events. These questions determine why something happened and what led to an outcome, which means deciding whether a change in one variable leads to a change in another variable. For example, what diet increases the sportsperson's stamina? Why did a particular sportsperson not develop stamina with the same diet?

Mechanistic

Mechanistic questions are used to understand the process of the underlying data. These questions will uncover details of the preceding questions to identify patterns, trends, etc., to make predictions about future behavior. The questions can be what the relationship between diet and sportsperson is, and so on. These question types help you to achieve a more profound understanding of how it works.

Characteristics of a Question

There are a few characteristics to consider when formulating a question. Mainly, the question should be meaningful and understandable.

The characteristics of a question are as follows:

- The questions are essential because the public data is accessible to all; someone might have prepared well and answered the same question. This helps you to reduce duplicity.

- The question should be a clear outcome that can be measured.

- The question should result in actionable insight that can inform decision making.

- The question must be answerable. Some good questions might not get an answer because specific data is not available.

- A hypothesis about the relationship between the variable and data must drive the question.

- The question must be self-explanatory, avoiding ambiguity and vagueness.

- The question must be relevant to the data available.

Linking Question to Data

When preparing questions, consider how the framed question translates into a data problem and what output it should generate. For example, your output may be whether taking a Vitamin B12 supplement is associated with muscle pain, but your question is about use of a pain reliever on every occasion.

Another type of issue concerns using appropriate data that leads to a good result. Let's take the example of the Covid pandemic. Imagine you are using a data set created from a survey of children who had Covid. The survey includes information about whether the children had heart problems that developed after their Covid vaccination, and you might see an association between the vaccine and expansion of the heart. However, generalizing all vaccine-related heart problems is inaccurate, which recalls the bias.

A good rule of thumb is if you are examining the relationship between two factors (vaccination and heart problem), bias might be a problem if you were more likely to observe individuals with certain factors because of the country, population, or vaccination type you selected.

The best way to link the question to data is to think about how you answer the question that you framed and who the audience of these answers and questions is. This will help you uncover insights and trends in the data, and the data itself provides a fundamental insight upon which these questions are based.

Prepare Cleansed Data

Cleansing data (gold data) is the process of getting the data into a format that is easily accessible. The availability of cleansed data is essential to data analysis. When collecting data, choose data relevant to the question you identified and representative of the population in which you are interested.

More than 80 percent of big enterprises today deal with unstructured and structured data scattered across ecosystems and globally available. The transaction systems generate structured data. These data only require a few rules to be cleansed, but unstructured data is generated globally on diverse platforms, like social media, medical devices, IoT, and so forth.

In the majority of cases, raw data related to unstructured data records such as X-ray, MRI images, and pathology reports are associated with the patient record in a hospital information system, however,these raw data cannot be used as-is for data analysis purposes. These data requires a pre-processing and transformation before they can effectively used for data analysis. The raw data cannot be used directly because of data format, quality of data, and privacy and security etc. Using raw data for analysis will slow the analysis process since the data scientist will have to do a detailed study to identify the right data sets.

When transforming raw data into cleansed data for data analysis, it is crucial to analyze and identify the essential data source based on the specific question or objective of the analysis. The raw data are in different forms like web pages, chats, video files, audio files, documents, emails, comments, and so on. It is essential to consider only those raw data that are relevant to you.

You need to spend a considerable amount of time and effort on cleaning the raw data for analysis. Hadley Wickham (Chief Scientist at RStudio) laid out the general principles for cleansing data for analysis, as follows:

- Each variable you analyze must be in one column.

- Each observation of the variable should be in a different row.

- There should be one table for each kind of variable.

- If you have multiple tables, they should include a column that allows them to be linked.

Along with the preceding principles, you need to take care of the following steps:

- First, understand the data analysis result. If you need clarification about the result, the cleansed data might not be helpful. It is essential to understand what outcome is required. Is it descriptive, causal, predictive, inferential, mechanistic?

- Second, determine the technology needed for data processing and storage. Even though the data come from different sources, the cleansed data is injected and stored in the technology stack for analysis. The choice of technology is purely dependent on the volume, scalability, velocity, and variety of other non-functional requirements. Some of the few technology stacks are Apache Hadoop, Cloudera, data bricks, snowflakes, etc.

- Third, split the training and testing data. The training data are used to train the model using various algorithms based on the question, and testing data are used to evaluate the model's performance.

Identify a Relevant Algorithm

This section helps you to understand the purpose of the model concerning identified data sets. Many models exist for data analysis, which can be used in combination to arrive at the desired results. They are as follows:

- **Statistical Models**: Statistical models are used to analyze data and make predictions based on statistical theory, such as linear regression, logistic regression, time series analysis, etc.

- **Machine Learning Models**: The ML models are used to build a predictive analysis for data by using various algorithms that automatically learn patterns by using the training set. The essential and most common models are neural networks, decision trees, support vector machines, random forests, etc.

- **Simulation Models**: These models are used to simulate real-world situations to help understand and predict the behavior of a system. These models are helpful in digital twin analysis. The most common models are agent-based models, Monte Carlo simulation, and discrete event simulation.

- **Cognitive Models**: These models are used to analyze the behavior of the human brain as far as how people think, learn, and make decisions. These models are commonly used in neuroscience, psychological research, and other AI research. The most common cognitive models used in are deep learning, neural network, natural language processing (NLP), expert systems, etc.

Choosing a suitable model is one of the essential steps in data analysis. Here are a few guidelines you may require when selecting the appropriate model for your analysis:

- Before choosing any model, you must clearly understand the problem and the objective.

- It is essential to understand the data sets, including data distribution, data sources, data variables, metadata, etc.

145

- It is essential to understand the assumption in each model. For example, linear regression assumes that the relationship between the independent and dependent variables is linear.

- Always keep performance in mind, so evaluate the model performance on the selected data set using various methods, such as accuracy, precision, recall, F1 score, etc. If your selected model is not up to the performance you expected, try to refine the model by changing the hyperparameters by modifying the features, or choose a different analysis model.

Build a Statistical Model

Generally, a model is something we build to understand the real world. A typical example is the use of a 3D image that mimics the organs of a human body to help us to understand the exact functioning of the organs. The same concept applies to data sets as you use the data to understand the real world.

In the consumer goods world, a marketing analyst has a data set collected from various sources, and the analyst uses the sample to predict the sales of their product against those of the competitor. The analyst uses the sample data to construct a model to predict what will be used for their product sales. Building a model involves imposing a specific structure on the identified data and creating a summary of the data. In this example, you may have hundreds of observations, so the models are a mathematical equation that reflects the pattern of the data. You use hundreds of observations for the summary.

As mentioned, a statistical model such as linear regression, logistic regression, and time series models provides a quantitative summary of your data and imposes the specific structure on the people from which data samples are sourced.

Let's consider an example of a survey of consumer goods. Assume you are in the field to ask various consumers about product choice, the price they want to pay, and the importance of the product for them, etc. The goal of this survey is to identify the product's popularity and relevance. You will not ask the entire country population; you may choose samples from every major and minor city. The following are the responses of 30 people from the town who volunteered the information:

10, 5, 20, 12, 8, 20, 25, 19, 18, 17, 21, 10, 3, 7, 12, 15, 22, 16, 8, 9, 12, 15,11, 10, 2, 7, 6, 9, 14, 20

The data is the actual price volunteers are willing to pay for a product, irrespective of the brand. The exact answer for the survey lies in the preceding data. You don't need bulk data to identify the summary of your question, as precise data is enough. This is precisely the statistical model's key element (data reduction). The process of data reduction is statistics. You can generate a mean or average of data to determine the number. Still, the statistical model is not just mean or average but requires some structure on the data, such as relationships between various data set factors.

Gaussian Model or Normal Model

The more popular statistical model is the Gaussian model, or normal model, which is commonly used to describe data distribution in a bell-curved shape. The Gaussian model's fundamental properties are a mean with a bell-curved shape and the standard deviation as the curve's width. The area under the curve is always equal to 1, which means the total probability of the outcome is equal to 1.

Let's take the product survey data—the product price that people are willing to pay. If I apply a Gaussian model to the data, the mean is $12.77, and the standard deviation is $5.98. The model looks like that in Figure 5-3.

This model (Figure 5-3) is generated with R Studio by using the following R function:

```
Mean = $12.77
Standard Deviation = $5.98

> Price<-seq(-10,40,by=0.1)
> population=dnorm(Price, 12.77, 5.98)
> plot(Price, population)
```

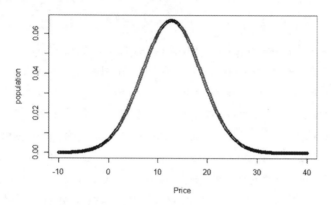

Figure 5-3. *Gaussian model for price*

You can use the Gaussian model to explore more questions on this. For example, if you want to find out the number of people who would like to pay $15 for your product, you can use this R function:

```
pnorm(15, mean = mean(x), sd = sd(x), lower.tail = FALSE)
Result = [1] 0.5
```

So, about 50 percent of the population would be willing to pay more than $15 for your product. You can use various combinations based on your question to identify the result. One thing you need to keep in mind is that data usually does not appear in the model, but I am showing the number to explain the use of a model.

Now, let's try a histogram representation of the same data to identify the exact population willing to pay each price.

A histogram is a graphical representation of a data set, and it displays the number of occurrences of data sets in a specific distribution. The histogram's x-axis represents the range of data values, while the y-axis represents the frequency or count of the data points. In this case, the y-axis represents the population, and the x-axis represents the price of the product willing to pay (Figure 5-4).

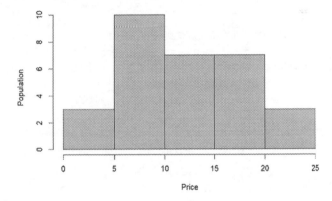

Figure 5-4. *Histogram*

I have taken all the samples to represent the histogram in the Figure 5-4 histogram. In this model, you can find what the population is willing to pay.

Linear Regression Model

Linear regression is one of the statistical models and helps to model the relationship between dependent variables in a predictive way. The objective of this model is to find the best-fit line that represents the linear relationship between variables. One variable is an independent variable (age), and another is a dependent variable (savings). The assumption is that the data sets are linearly related, so the regression model is used to

149

identify a linear function to predict the response value of the dependent variable—as accurately as possible—as a function of the feature or independent variable.

Let's consider a simple example of age group and respective savings—the savings are linearly related to the person's experience. The data set is a linear relation between experience/age group and savings. Age group doesn't need to be dependent on experience and salary, but some critical relationship exists. Table 5-1 shows the data set to identify a linear regression.

Table 5-1. *Age Group and Savings Data Set*

Age	savings in $	Age	savings in $	Age	savings in $
22	4000	37	10450	52	15789
23	4344	38	10456	53	15989
24	5789	39	10789	54	16120
25	6388	40	11000	55	16345
26	6556	41	11345	56	16455
27	6856	42	11456	57	16989
28	7178	43	12039	58	17298
29	7389	44	12569	59	17989
30	7765	45	12989	60	18500
31	7959	46	13100		
32	8367	47	13250		
33	8756	48	13909		
34	9469	49	14568		
35	9658	50	15000		
36	10278	51	15345		

For the aforementioned sample, I have stored it in the Savings_Data.txt file. After loading the data into RStudio, first you should check that the data has been read correctly by using summary().

summary(Savings_Data)

Both the variables (age and savings) arc quantitative. When you run this function, you can see in Figure 5-5 the dependent and independent variables and a numeric summary.

```
     Age            Savings
 Min.   :22.0   Min.   : 4000
 1st Qu.:31.5   1st Qu.: 8163
 Median :41.0   Median :11345
 Mean   :41.0   Mean   :11551
 3rd Qu.:50.5   3rd Qu.:15172
 Max.   :60.0   Max.   :18500
```

Figure 5-5. *Dependent and independent variable summary*

In this example, I have only one dependent and independent variable and don't require testing of any other variables.

- First, let's check the histogram for age and savings. Figure 5-6 is how it looks.

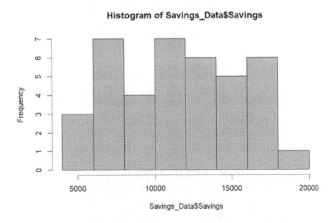

Figure 5-6. *Histogram age and savings*

151

The observation of the histogram is roughly bell-shaped. Now we can move to a linear regression chart.

- Secondly, let's check the linearity of the data by using a plot.

```
plot(Age ~ Savings, data=Savings_Data)
```

In Figure 5-7 the plot shows that the relationship between age and savings is linear.

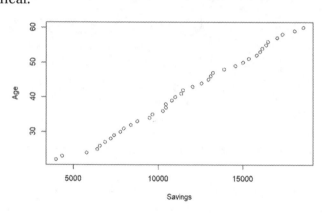

Figure 5-7. *Relationship between age and savings*

I have checked the data for linearity, and now I am conducting the linear regression analysis on the data (age and savings).

```
savings.age.sa<-lm(Age~Savings, data=Savings_Data)
summary(savings.age.sa)
```

The following table represents the model equation and summarizes the model residual.

```
all:
lm(formula = Age ~ Savings, data = Savings_Data)
```

Residuals:

Min	1Q	Median	3Q	Max
-1.44479	-0.68372	-0.05374	0.56718	2.31748

```
Coefficients:
            Estimate Std. Error t value Pr(>|t|)
(Intercept) 8.436e+00  4.430e-01   19.05    <2e-16 ***
Savings     2.819e-03  3.626e-05   77.75    <2e-16 ***
---
Signif. codes:  0 '***' 0.001 '**' 0.01 '*' 0.05 '.' 0.1 ' ' 1

Residual standard error: 0.9012 on 37 degrees of freedom
Multiple R-squared:  0.9939,    Adjusted R-squared:  0.9938
F-statistic:  6045 on 1 and 37 DF,  p-value: < 2.2e-16
```

The preceding details represent the model equation and summarize the model residual, and the coefficients section provides the following details:

- The estimate for the model parameters is as follows: the value of the intercept is 8.436e+00 (8.436), and the estimated effect on savings is 2.819e-03 (0.002819) (Note: *8.436e+00 is scientific notation for the number 8.436, e+00 means to move the decimal point 0 places to the right and same e-00 means to move the decimal point three places to the left*).

- The second is the standard error, which means how much the sample mean varies if you repeat the same study with different data sets. Here the intercept is 4.4303e_01(0.44303).

- The third is the t-value (test statistics), which is a number calculated from a statistical test hypothesis and helps to calculate the p-value. The t-value here is 19.05.

- The fourth is the p-value, the probability of finding a given t-value if the null hypothesis were true. Here the p-value is <2e-16 (0.0000000000000002).

153

From the preceding result, you can conclude that there is a significant positive relationship between age and savings (p-value is < 0.0001).

The code produced the graph shown in Figure 5-8.

Residual vs. Fitted

The residual and fitted value refers to the difference between the actual value of the dependent variable (age) and the predicted values of the age based on the regression equation. In other words, the fitted values are the predicted values of the dependent variables based on the independent variables, and the residual is the differences between the observed values and predicted values (Figure 5-8). This is useful for finding the best fit of the regression model.

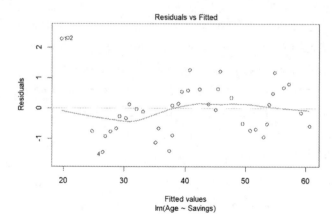

Figure 5-8. *Fitted vs. residuals on age/savings*

In Figure 5-8, the residual values are randomly distributed around zero with no pattern or trend, indicating that the model is a good fit for the data. If the residual has a pattern like a U shape or a curve or some other shape, it would indicate that the model does not capture all of the relevant information of the sample data.

Normal Q-Q (Quantile-Quantile)

The Q-Q helps you to check the normality assumption of the data set. The following plot compares the standard data to theoretical quantities by plotting the ordered data set against the expected value of a standard normal distribution.

In the Figure 5-9 plot, the data resides in a straight line, suggesting the data is usually distributed. If the data deviate from the straight line, then it indicates the data is not normally distributed.

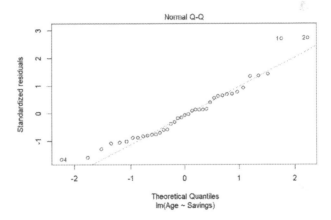

Figure 5-9. *Normal Q-Q of age/savings*

If the data deviated upward or downward from the straight line, this would indicate skewness in the data. If the data deviated more in the tails than in the middle, this would indicate the thin-tailedness of the data.

Scale Location

The scale location helps you to check the assumption of equal variance of the errors. The plot provides the square root of the standardized residual versus fitted values of the response variable. The standardized residuals are divided by the estimated standard deviation and transformed by taking the square root to stabilize the variance.

In Figure 5-10, data points are randomly scattered around a horizontal line, which helps you to identify whether the variance is reasonable and whether errors are in the same variance. In the diagram, the data points are formed in a funnel shape as shown in Figure 5-10, which indicates that the errors have different variances that change across the range of the independent variable. As you can see, the funnel shape is widened. This indicates that the variance of the errors increases with the level of an independent variable.

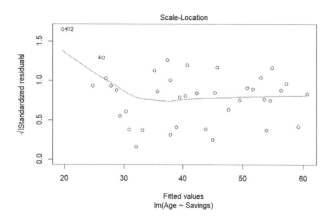

Figure 5-10. *Scale location of age/savings*

Residual vs. Leverage

The residual versus leverage helps you to identify those observations that may affect the regression results. The leverage measures the outlying quantum of observation in terms of predictor value, as shown in Figure 5-11.

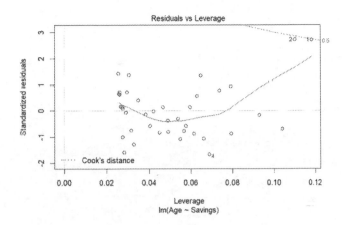

Figure 5-11. *Residual vs. leverage of age/savings*

If an observation has a high leverage value and a large residual, this indicates the observation may be an influential outlier that affects the regression result.

After all the analyses, plot the data and the regression line from the linear regression model.

The first step is to plot the data points on a graph by using the following lines of code:

```
> saving.graph<-ggplot(Savings_Data, aes(x=Savings,
y=Age))+geom_point()
> saving. graph
```

Figure 5-12 shows what the graph looks like without the regression line.

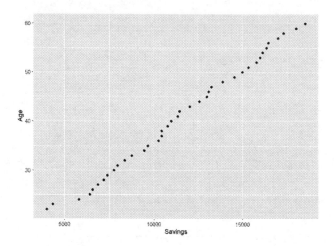

Figure 5-12. *Regression line plot*

With the regression line, add a geom_smooth and lm to create a line, as shown in Figure 5-13.

```
> saving.graph<- saving.graph + geom.smooth(method = "lm",
col="black")
> saving.graph
```

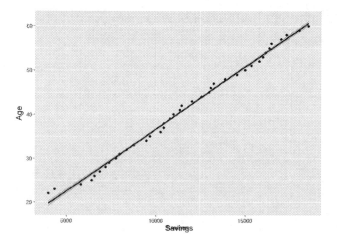
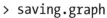

Figure 5-13. *Regression with geom_smooth*

The regression line represents the relationship between age and savings. It was used to predict the value of savings based on age and the change in the dependent variable for every one-unit increase in the independent variable. The regression line is also known as the best-fit line.

When we stop the iteration of the data analysis process, the answer is dependent on the data sets and results you expected. Usually, it would be best to run multiple iterations to cross-verify the data and plot to arrive at a final conclusion. It is essential to understand that the data analysis you are conducting has a conclusion. It would be best if you made a specific decision based on your audience and who you are presenting.

Match Result

Matching the result of a model involves understanding what the output means in the context of the problem being solved and the type of data used to train and test the model. Matching the result of a classification model involves understanding the predicted class label assigned to each data set. This could involve looking at the possibilities assigned to each class model. It is essential to understand the following evaluation matrices to assess the quality of the classification result:

- **Recall**: This is an evaluation metric that measures the fraction of valid positive instances among all actual positive instances. *Recall = True Positives/ (True Positives (TP) + False Negatives (FN))*. TP = Number of instances that are correctly predicted as positive, FN = Number of instances that are incorrectly predicted as negative.

- **Precision**: This is an evaluation metric that measures the actual positive instances among all predicted positive instances. *Precision = True Positive / (True Positive (TP)+ False Positive (FP))*. TP = Number of instances that are correctly predicted as positive, and FP = Number of instances that are incorrectly predicted as positive.

- **F1-score**: This is an evaluation metric that is the mean of the precision and recall evaluation metrics. *F1-score = 2*(precision * recall)/ (precision + recall)*. The score ranges from 0 to 1. A 1 indicates good precision and recall metrics, so the model is performing well, and a 0 indicates bad precision and recall metrics, so the model is not performing.

- **Confusion Matrix**: This is commonly used to evaluate the performance of an ML model in a classification problem. The matrix shows the number of true positives, true negatives, false positives, and false negatives that the model produces. In a binary classification problem, the confusion matrix is a 2x2 table with four values. The four values are True Positive (TP), False Positive (FP), False Negative (FN), and True Negative (TN).

In the statistical model, you need to understand the estimate of the model parameters and the associated statistical significance of those estimates. As shown in the previous section, the statistical model aims to estimate the relationship between one or more independent variables and dependent variables.

In the linear regression model, the objective is to estimate the relationship between the independent variable and the dependent variable by fitting a linear equation to the data, as shown in Figure 5-8. The coefficient of the linear equation is shown in the Normal Q-Q model, which involves understanding the direction and strength of the relationship between the dependent and independent variables. In addition to the Residual vs. Fitted and Normal Q-Q models, the p-value helps you to determine whether the estimated parameters are likely to be different from zero or not.

Matching the statistical model of the estimated parameters involves comparing the p-value intervals to the pre-defined significance level, usually set at 0.05 or 0.01. Suppose the p-value is less than the significance level. In that case, the estimated parameters are statistically significant, and if they are more remarkable than the significance level, then the estimated parameter is not statistically significant.

Create an Analysis Report

When creating an analysis report, you need to follow several key steps, but there are no universal rules dictating that the report should look like this or like that. This section provides a few valuable tips you should consider when preparing the report.

Introduction

Write a readme file or text about the project, the objective and type of questions you have selected to address, sample size, and data source. All this text will provide an abstract view of everything you have done in your data analysis.

Define the Question

Only some analyses start with a good question. Still, it would be best if you put a little effort into identifying the right question, because defining an objective-based question is a critical step. If you can normalize your question as precisely as possible, you will reduce the noise during the analysis of the data set. The question must be specific, measurable, and relevant to the research analysis. With well-formulated questions, your report can provide insights and inform decision making.

161

Data Preparation

The data is an essential factor in an analysis report. The report must include the data source, data collection process, and relevant details about the data, like metadata, sample size, and so on, as described in Chapter 4, "Datafication Pipeline." You need to include the data preparation process, data cleansing process, data transformation, and so on. The details of data collection are crucial because they explain the quality of the data.

In the report, you must include the figures to show the data sources and cleansing methods you put together to look at the data produced.

Statistical Model Details

The statistical model details are essential in the report, which describes the methods and procedures you used to analyze the data. In this, you need to define the model you selected and why you did so. Explain assumptions and limitations, and explain the exploratory data analysis you conducted before applying the model.

Identify the dependent and independent variables used in the model and define the measurement scale of these variables. Provide estimates of the model's parameters, including residual versus fitted, coefficient, p-value, and confidence intervals with specifications.

All these details must be present in the statistical measure in a clear and organized manner, using tables, graphs, and charts. It is also essential to match the statistical result in the context of the question you identified and the solution you identified, with limitations and assumptions.

Diagnosis Result

Explain the details of the assessment you have conducted to apply the statistical model against assumptions like the normality of residual, independence of errors, etc. Generate the diagnosis plot against the assumption by using R.

Conclusion and Recommendation

Summarize the fundamental analysis and provide a recommendation if needed for further analysis, and include references. Use R Markdown to generate a report.

Present the Report

Prepare the report and organize the content based on the analysis. Communicate clearly and concisely to the audience.

Summary

This chapter provides an insightful approach to doing data analysis. However, each data set has its unique needs when doing an analysis, including issues and assumptions. This chapter provides the framework for developing a question, exploring your data, using a suitable model for the data, and providing data analysis reports. This framework is standard across the types of data analysis you conduct.

Conducting a thorough analysis at the outset of an analytics project is crucial to its success. It helps to ensure that the analysis is well-planned, targeted, and aligned with the desired outcoes due to clarity of objectives, data identification, data quality, resource planning, method selection and risk analysis, you need to have a through analysis is essential. It is essential to define a question you want to answer as this will help guide the entire data analysis process. It is important to understand the purpose of the analysis and what insights you hope to gain from it. Finally, establishing clear goals and objectives will help ensure that the analysis is focused and directed toward achieving specific outcomes.

CHAPTER 6

Sentiment Analysis

Sentiment analysis is a process used in natural language processing (NLP) to analyze and extract information related to individual opinions and emotions and determine the information's polarity, such as positive, negative, or neutral. The analysis process involves several steps: data collection, pre-processing, feature extraction, classification, and evaluation.

Sentiment analysis can be used to analyze customer feedback, predict stock market trends, estimate success of new product launches, analyze elections, and so on. This analysis helps businesses and political parties to make data-driven decisions.

In this chapter, I will provide high-level details of sentiment analysis and the algorithm used. It is impossible to cover everything in one chapter, but this helps you get a foundation on which to construct the sentiment analysis tower.

You can find answers to the following:

- Usage of sentiment analysis

- What are the types of sentiment analysis?

- What are the different ways to pre-process the data?

- What are the techniques to be used for analyzing the sentiment?

© Shivakumar R. Goniwada 2023
S. R. Goniwada, *Introduction to Datafication*,
https://doi.org/10.1007/978-1-4842-9496-3_6

Introduction to Sentiment Analysis

Sentiment analysis is the computational identification and categorization of opinions expressed in a text, especially to determine whether the writer's attitude toward a particular topic, product, person, or other entity is positive, negative, or neutral. It is widely used in social media monitoring, customer feedback, product reviews, market research, and political analysis.

With the high level of social media usage, sentiment analysis has become increasingly critical for businesses, organizations, and political parties to understand customer and voter sentiment and preferences for data-driven decisions made by executive boards.

The methods and technologies applied can include natural language processing, text analytics, computational linguistics, and even event biometric data. These are used to systematically identify, extract, quantify, and study affective states and subjective information.

There are several approaches to sentiment analysis, including the following:

- **Rule-Based Process**: It uses a set of pre-defined rules and lexicons to identify sentiment in text data.

- **Machine Learning**: It uses naïve Bayes, support vector machines (SVM), and decision trees trained on labeled data to classify sentiments.

- **Deep Learning**: It uses convolutional neural networks and recurrent neural networks for complex data in social media.

Use of Sentiment Analysis

Sentiment analysis can be used in a wide range of industries and domains, including the following:

- **Voice of the Customer**: It is widely applied to the voice of customer actions, such as reviews, survey responses, and social media comments.

- **Internal Threat Detection**: The enterprises may monitor their workforce communication to detect potential information security threats, reduce the divulgence of sensitive information, or protect intellectual property.

- **Product or Brand Reviews**: Industries, especially consumer goods industries, use sentiment analysis to monitor social media platforms to understand how consumers perceive their brand, issues of products, customer expressions, price range, etc.

- **Product Feedback**: It analyzes consumer feedback through surveys, reviews, and feedback to understand consumer thinking.

- **New Product Launch**: Sentiment analysis can be used in market research to analyze public opinions to decide on potential opportunities for new products.

- **Election in Democracies**: It is used by political parties to understand voters' opinions of political candidates, policies, leaders, and issues.

- **Financial Analysis**: It is used to analyze new articles, news, and investor sentiments to identify trends and predict market progress or movement.

Types of Sentiment Analysis

Various types of sentiment analysis can be used to analyze text data. We will look at a few types.

Document-Level Sentiment Analysis

Document-level sentiment analysis analyzes text, such as articles, social media posts, blogs, and so on and determines whether the text in the document has a negative or positive sentiment.

Let's consider an example of a document with text:

> "I bought a car, Honda City model 2023, today and WOW, what a difference compared to my earlier lower end of the car, it boost my self respect. When I bought the car at home, It was tricky to park in my car shed was sparse because it was not used for some time, but the helper cleaned the entire area within a few hours. I am a pleased car owner, and it will benefit my kid to go to school. It is outstanding consumers perceive their brand's hope that we get the great feeling all over the year!"

For the aforementioned text, I will apply the sentiment classifier, document encoder, and sentiment encoder.

Figure 6-1. *Analysis with sentiment dictionary*

By using the dictionary as shown in Figure 6-1, you can provide a sentiment score for each piece of text that has been bolded in the following passage:

> "I bought a car, Honda City model 202,3 today and **WOW**, what a difference compared to my earlier lower end of the car, it **boost** my self respect. When I bought the car at home, It was **tricky** to park in my car shed was **sparse** because it was not used for some time, but the helper **cleaned** the entire area within a few hours. I am a **pleased** car owner, and it will **benefit** my kid to go to school. It is **outstanding** consumers perceive their brand's **hope** that we get the **great** feeling all over the year!"

You can calculate overall sentiment score using Table 6-1.

Table 6-1. *Sentiment Analysis Scoring*

Positive Sentiment		Negative Sentiment	
WOW	1.0	Tricky	-0.4
Boost	0.6	Sparse	-0.2
Pleased	0.2		
outstanding	0.8		
Hope	0.2		
Great	0.5		

The sentiment score is as follows:

Overall Sentiment = +3.3 -0.6=+2.7, is "positive"

Note   These scores are based on a publicly available dictionary.

Aspect-Based Sentiment Analysis

Aspect-based sentiment analysis helps identify the sentiment of specific product or service aspects. This enables you to understand the behavior of a customer's perception of a product. The outcome can help you to make the right decisions. This can be used for product reviews, customer feedback analysis, market research, and so forth.

The aspect-based process involves these three steps:

- **Feature Extraction**: In this step, you will extract the feature of the product or services from the data. The extraction is done using dependency parsing, rule-based methods, or part-of-speech tagging.

- **Sentiment Classifier**: Once the features are identified, identify the sentiment polarity of each segment in terms of positive, negative, or neutral. This can be done by using the sentiment lexicon or dictionary.

- **Summarization**: Summarize the features and polarity and provide an overall sentiment score for the product.

Let's take an example of a restaurant review. I have downloaded the data set from Kaggle.com,[1] and you can use the following code to formulate an aspect-based sentiment analysis.

```
library(tidytext)
library(dplyr)
reviews <- read.csv (//Restaurant_Reviews.csv",
stringsAsFactors = FALSE)
# load AFINN sentiment lexicon
afinn <- get_sentiments("afinn")
# define aspect categories
```

---

[1] https://www.kaggle.com/datasets/cristeaioan/ffml-dataset

```
aspect_categories <- tibble(aspect = c("food", "service",
"ambiance"), keywords = list(c("food", "menu",
"dish", "cuisine"),  c("service", "waiter", "staff",
"manager"),  c("ambiance", "atmosphere", "decor")))
# unnest review text into individual words
words <- reviews %>%  unnest_tokens(word, Review)
%>%  filter(!word %in% stop_words$word)
# calculate sentiment for each aspect category and review
aspect_sentiment <- aspect_categories %>%  unnest(keywords)
%>%  inner_join(words, by = "word") %>%
  group_by(Review_ID, aspect) %>%
  summarize(sentiment = ifelse(all(keywords %in%
  unique(words$word)), sum(afinn$value[afinn$word %in%
  unique(words$word)]) / n(),  NA_real_))
```

Multilingual Sentiment Analysis

Multilingual sentiment analysis is a process of analyzing sentiment in multiple languages; this is a complex sentiment analysis as each language has its own structures, vocabularies, single-byte language, multi-byte language, and culture nuances that impact the sentiment analysis.

This sentiment analysis can be performed using lexicon-based machine-learning and deep-learning techniques.

In the following example, I identify a sentiment score using 'textcat' in the R programming language.

```
install.packages("textcat")
library(textcat)
text <- c("Je suis de Paris et j'habite à Washington",    "Ich
komme aus Bengaluru und Moro in Dubai",   "Sunt din Dublin
și locuiesc în Singapore.",    "Vengo da Mosca e vivo a
```

```
Berlino")    #I am from Paris and living in Washington -
French, #I am from Bengaluru and living in Dubai - German, #I
am from Dubli and living in Singapore - Romania, #I am from
Moscow and living in Berlin - Italy
multilingual <- textcat(text)
lingual_sentiment <- lapply(multilingual, function(x)  {
  if(x == "fr") {
    data("afinn_fr")
    score <- sum(afinn_fr[unlist(strsplit(text[x], " "))])/
    length(text[x])    }
  else if(x == "de") {
    data("afinn_de")
    score <- sum(afinn_de[unlist(strsplit(text[x], " "))])/
    length(text[x])    }
  else if(x == "es") {
    data("afinn_es")
    score <- sum(afinn_es[unlist(strsplit(text[x], " "))])/
    length(text[x])    }
  else if(x == "it") {
    data("afinn_it")
    score <- sum(afinn_it[unlist(strsplit(text[x], " "))])/
    length(text[x])    }
  return(score)
}
result <- data.frame(text, lang, lingual_sentiment)
print(result)
```

The following table are the sentiment scores of the multilingual analysis:

```
text                                                Language
 WordCount SentimentGI NegativityGI

1 Je suis de Paris et j'habite à Washington     french
  48         0.125        0.1041667
2      Eu sou de Bengaluru und Moro in Dubai    german
  48         0.125        0.1041667
3  Sunt din Dublin și locuiesc în Singapore     romanian
  48         0.125        0.1041667
4           Vengo da Mosca e vivo a Berlino      Italian
  48         0.125        0.1041667
```

Pros and Cons of Sentiment Analysis

Recently, sentiment analysis has grown, and most decisions are taken in corporate, electroal politics etc. are based on sentiment calculation. The few significant challenges of sentiment analysis are as follows:

- **Privacy Concern**: This is one of the primary concerns as sentiment analysis uses text data to analyze the text, which can raise privacy concerns. Organizations must ensure that they have data protection to safeguard user data. A few data protection laws are GDPR, HIPAA, and DPDP from India.

- **Emotion Identification**: The outcome of sentiment analysis is either positive, negative, or neutral, and most of the time, it fails to identify humans' emotions. To identify emotions, you need a specialized model.

- **Domain Analysis**: Sentiment analysis does not differentiate the domain-specific analysis, such as health care, e-commerce, energy, etc. Each domain has its language to identify the texts, and the sentiment models may not be able to identify them accurately.

- **Multilingual Analysis**: It is challenging to analyze multilingual texts, as the model requires training in each language.

- **Context Analysis**: The sentiment models struggle to identify the context. A few texts differ based on the context.

Pre-Processing of Data

Data pre-processing is a significant and crucial step for any sentiment analysis. There are various techniques available to pre-process data. Let's look at a few techniques.

Tokenization

Tokenization breaks down a text into individual tokens, such as words, phrases, or sentences. It is an essential step in sentiment analysis. The tokenization can be done using various techniques depending on the nature of the task and requirements. The following are a few techniques:

- **Word Tokenization**: Word tokenization helps you to break down the provided sentences into each word. For example, "**I interested in the Samsung UHD Television**". This sentence tokenized into the words "*I*", "*interested*", "*in*", "*the*", "*Samsung*", "*UHD*", and "*Television*".

- **Sentence Tokenization**: Sentence tokenization helps you to break down a text into individual sentences. For example: "**I love dogs. My wife also loves dogs.**" This sentence tokenized into *"I love dogs"* and *"My wife also loves dogs."*

- **N-gram Tokenization**: N-gram tokenization helps you to break down a sentence into an immediate sequence of *n* words. For example, "I will be in Bangalore during the month of April." This sentence tokenized into *"I will", "will be", "be in", "in Bangalore", "Bangalore during", "during the", "the month", "month April".*

Stop Words Removal

Stop word removal helps you to remove common words like "the", "a", "and", "the", etc. These words do not carry any meaning for sentiment analysis. Removing these words helps you to reduce noise in the data set, but be cautious about the removal of stop words because they do carry some contextual information.

Stemming and Lemmatization

Stemming and lemmatization are a natural language processing technique to reduce words to their base or root form. This improves the performance of the analysis by reducing the number of distinct words.

Stemming is a process of removing the suffix from a word to obtain its base form—the stem. It uses the heuristic technique to operate by applying rules to remove common suffixes, like "-ing", "-ed", "-r", etc.; for example, "eating" would be "eat".

```
> library(SnowballC)
> word <- "eating"
> stem_word <- wordStem(word, language = "english")
> print(stem_word)
[1] "eat"
```

Lemmatization is the process of reducing a word to its base form, called the lemma, by considering the part of speech and the context of the word. It uses the dictionary to look up the base form of the word. For example, the lemma of "catching", "catcher", and "catch" would be "catch".

```
library(udpipe)
model <- udpipe_download_model(language = "english")
ud_model <- udpipe_load_model(file = model$file_model)

word <- "catching"
doc <- udpipe_annotate(ud_model, x = word)
lemma_word <- doc$lemma[doc$token_id == 1]
print(lemma_word)
output: catch
```

Handling Negation and Sarcasm

Handling negation and sarcasm requires two standard linguistic processes that can affect the meaning of the text and present challenges in natural language processing.

Negation refers to the use of negative words to negate the meaning of a sentence, such as "not", "no", "don't", etc. For example, "**I do not want to go to Bangalore**" negates the positive sentiments of "I want to go to Bangalore." The most important question is how to handle a negation in NLP. The answer is to use double negation cancellation, which involves detecting double negations and canceling them out to determine the true sentiment of the sentence. For example, "I do not want to go to

Bangalore" can be interpreted as a positive sentiment by using sentiment shifters such as "but", "although", "because", etc. Adding these shifters will provide a positive sentiment approach; for example, "I do not want to go to Bangalore, because I have a job offer in my hometown."

You can use lexicons for negation handling by using rules and annotations, such as negation words' being assigned negative weights. This helps to evaluate the given context correctly.

Sarcasm is words or phrases that convey the opposite meaning of what is mentioned. For example, "Oh, super, we wanted to go to the theme park" in this example, "we wanted to go to" is sarcasm because it gives the opposite sentiment of what the person thinks. The person is likelier to think, "do not want to go park". The tone and situation also play a role in conveying the message.

Negation and sarcasm require specialized techniques and a deep understanding of languages.

Rule-Based Sentiment Analysis

Rule-based analysis uses pre-defined rules to identify sentiment in text or phrases. The rules are determined based on the linguistic and grammatical features of positive, negative, and neutral emotions.

The steps involved in rule base sentiment analysis are as follows:

- **Text Processing**: Process text by removing unwanted words and reducing noise.

- **Lexicon Creation**: Create a sentiment lexicon, which is a list of words that are associated with a specific sentiment (positive, negative, and neutral).

- **Scoring**: Score each word after lexicon creation based on negative, positive, and neutral.

- **Aggregation**: Aggregate each score to calculate the overall sentiment.

- **Classification**: Based on the overall sentiment score, you can identify whether it is positive, negative, or neutral.

For example, the first is to process the extraction of words, and then each of these extracted words is assigned a sentiment score based on the rules. For example, the word "great" might be assigned a positive sentiment score, the word "fall" might be assigned a negative sentiment score, and "good" might be a neutral score. Each sentiment score of all the words is aggregated to calculate the overall sentiment score of comments or reviews.

Lexicon-Based Approaches

The lexicon-based approach uses a list of words annotated with their corresponding sentiment designation as positive, negative, and neutral. A pre-defined sentiment lexicon or publicly available dictionary can be used. The lexicon-based approach involves matching the extracted words with the corresponding sentiment polarity in the lexicon and using this information to determine the overall score. There are two main types of lexicon-based approaches, as follows:

- **Corpus-based Approach**: In this approach, the lexicon is defined based on the occurrence of the word in a corpus text. The score of each word is based on the occurrences in positive or negative polarity.

- **Dictionary-based Approach**: This approach uses the pre-built publicly available lexicon to identify the sentiment polarity of each word. Then scores are aggregated to calculate the overall sentiment score.

Sentiment Dictionaries

It is commonplace to identify a lexicon in rule-based sentiment analysis. This dictionary stores the words annotated with positive, negative, and neutral polarity. During the process, you must invoke dictionaries to match the word and assign a score based on the polarity.

There are publicly available dictionaries, and you can use these based on the type of analysis you are conducting, as some are free and some paid. The following are a few dictionaries:

- **VADER**: It provides a combination of lexicons and grammatical rules to determine positive, negative, and neutral polarity scores.

- **SentiWordNet**: Most commonly used dictionary and provides a score based on their sense of positive, negative, and neutral polarity.

- **Opinion Lexicon**: This provides scores based on their sense of the positive, negative, and neutral polarity.

Pros and Cons of Rule-Based Approaches

Rule-based analysis has its advantages but may only be helpful in some contexts, especially when you require high accuracy and adaptability. Here are a few pros and cons of this approach:

- The pros lie in transparency because pre-built rules are used to determine the sentiment of a text, and it is swift as it uses these pre-built rules. The main pros are that it uses less training data and can customize for domains; for example, analyzing sentiment scores for health care, energy, finance, etc.

- The cons are that it is not providing enough accuracy for negations and sarcasm, and other processes also.

Machine Learning–Based Sentiment Analysis

Machine learning–based sentiment analysis is a natural language processing (NLP) technique that trains machine learning algorithms to classify text data into sentimental polarity, such as positive, negative, and neutral. If you have a large set of data, then this approach helps you to analyze it automatically. There are various approaches available for ML-based sentiment analysis, including supervised learning and unsupervised learning. Supervised learning training is a model based on labeled data, and unsupervised learning training is a model based on unlabeled data.

Supervised Learning Techniques

This sentiment analysis approach uses the data set labeled with positive, negative, and neutral polarities to train the model. The labeled data is pre-processed, features are extracted, and a relevant model is used and trained on the data set. Once the training is carried out, you need to check the model's performance. Once you complete these steps, you need to use the actual data to measure natural categories. The outcome of this model is based on the quality of the trained data set.

This model best suits social media, customer feedback, and market research purposes.

I have used a small set of lyrics review[2] data from Kaggle. Here is the sample code snippet using the naïve Bayes model:

```
install.packages("tm")
install.packages("caret")
library(tm)
```

---

[2] https://www.kaggle.com/datasets/kauvinlucas/30000-albums-aggregated-review-ratings

```
library(caret)
lyrics_data<-read.csv("labeled_lyrics_dataset.csv")
corpus <- VCorpus(VectorSource(lyrics_data$text))
corpus <- tm_map(corpus, removePunctuation)
corpus <- tm_map(corpus, content_transformer(tolower))

corpus <- tm_map(corpus, removeWords, stopwords("english"))
corpus <- tm_map(corpus, stemDocument)

dtm <- DocumentTermMatrix(corpus)
dtm <- removeSparseTerms(dtm, 0.9)

index <- createDataPartition(lyrics_data$sentiment, p = 0.6)

unique_sentiments <- unique(lyrics_data$sentiment)
print(unique_sentiments)

train_lyrics_data <- lyrics_data[index, ]
test_lyrics_data <- lyrics_data[-index, ]
model <- train(train_lyrics_data$text, train_lyrics_
data$sentiment, method = "nb", trControl = trainControl
(method = "cv", number = 10))
predictions <- predict(model, newdata = test_lyrics_data$text)
conf_matrix <-confusionMatrix(predictions, test_lyrics_
data$sentiment)
print(conf_matrix)
```

The output of confusionMatrix() will be a confusion matrix table that summarizes the TP (true positive), TN (true negative), FP (false positive), and FN (false negative) values for the predictions, as well as various performance metrics, such as accuracy, precision, recall, and F1 score (F1 score = 2*(precision* recall)/(precision+recall).

Unsupervised Learning Techniques

Unsupervised learning is a process of identifying a sentiment without using labels, and it depends on the existing structure to identify sentiment. It uses the clustering process to cluster the data set into different groups based on the similarities of the features. This can be done using k-means, hierarchical, or density-based clustering. It analyzes the sentiments based on their occurrences and the polarity of the texts in the cluster. The positive sentiment cluster has more positive texts, and the negative cluster has more negative texts, and it uses the lexicon libraries to identify the polarity.

Pros and Cons of the Machine Learning– Based Approach

Similar to a rule-based approach, the ML-based approach has pros and cons regarding the data set's quality, domain-specific analysis, and so on.

As you already know, the ML-based approach requires training the model before using it, so the accuracy of the data is critical. If you use a random sample from social media or reviews, the model will provide the wrong analysis. The ML approach is complex and requires a dedicated data science team and a high-processing system to train and execute a model.

A pro of the ML model approach is that the analysis can be carried out for a large data set and can scale to improve performance. Once you train the model accurately, it is easy to analyze for polarity.

There are other techniques, like **deep learning–based** sentiment analysis, that use deep neural networks to extract features for analysis. It requires training a deep neural network on a large labeled data set to identify patterns and relationships. This subject are not covered in this book.

Best Practices for Sentiment Analysis

As I explained in previous sections, sentiment analysis is the best way of analyzing data in datafication. Here are a few best practices you can adopt:

- Before analyzing sentiment and opinions, define the scope and what data you need to use for the analysis. The scope of analysis helps you to identify the best technique.

- If your scope requires the use of machine learning models and deep learning models, you need to train your model on labeled or unlabeled data, depending on model type.

- Use a validation data set to measure the accuracy by using the precision, recall, or F1 score

- Once you have analyzed the data, try to interpret the result for your goals and scope for trend analysis.

- There is no one-stop, or one evaluation that is sufficient to analyze the sentiment. It would be best if you improved the model continuously for better accuracy.

Summary

Sentiment analysis is a powerful technique for analyzing and understanding textual data. With the help of ML models and NLP algorithms, sentiment analysis can be used to automatically classify text as positive, negative, or neutral, providing insights into customer opinions, social media trends, and more.

There are various types of sentiment analysis, such as text-based, document-based, aspect-based, and multilingual analysis. Each type focuses on specific categories. You can use this chapter for the initial journey into sentiment analysis.

CHAPTER 7

Behavioral Analysis

Behavioral analysis studies behavior patterns to understand why humans act a certain way. It involves collecting and processing behavioral data, such as online activities, social interaction, social behaviors, social media activities, Internet of Things (IoT) sensors, and so on.

Behavioral analytics can be used in psychological and biological sciences to understand the patterns of behavior and identify the cause of mental issues, biological changes, and more.

In this chapter, I cover how data are collected for behavioral analysis, the importance of behavioral science, segmentation, details of behavioral analysis models, and how to perform behavioral analysis using descriptive and causal analytics with experimentation.

Introduction to Behavioral Analytics

Behavioral analytics is the process of understanding human behavior and machine behavior. The analysis includes analyzing social media, digital platforms, IoT sensors, and more.

Behavioral analytics is used to understand the behavior of each person. In this analysis, you need to understand why person A bought a particular item and why not person B did not buy it. We also want to understand why someone likes a particular locale to reside in and why someone doesn't want to live in a particular locale. This knowledge of data allows you to predict what type of people like things and why others do not.

© Shivakumar R. Goniwada 2023
S. R. Goniwada, *Introduction to Datafication*,
https://doi.org/10.1007/978-1-4842-9496-3_7

Human behavior is incredibly complex and influenced by a wide range of factors, making it challenging to analyze and predict accurately. While physical objects and systems often follow predictable patterns and behaviors, human beings possess consciousness, self-awareness, and the capacity for free will, which introduces a level of variability and unpredictability. The influences on human behavior are vast and interconnected. Social and cultural factors shape our beliefs, values, and norms, which in turn influence how we behave in different social contexts. Psychological factors such as personality traits, cognitive processes, and emotional states also play a significant role in determining behavior. Additionally, biological factors, including genetics, brain chemistry, and physiological processes, interact with social and psychological factors to influence human behavior. Moreover, human behavior is not static; it evolves and changes over time. Individuals learn, adapt, and grow through experiences, and their behaviors can be influenced by these evolving circumstances. This dynamic nature of human behavior further adds to its complexity and makes it challenging to predict with certainty. Therefore, you may need more than one type of analytics to determine the actual behavior. It would be best if you used a combination of many analytics, such as descriptive, predictive, and causal analytics.

As I mentioned in Chapter 5, "Data Analysis," there are four types of analytics: descriptive analytics, predictive analytics, prescriptive analytics, and causal analytics. Out of these four, mainly descriptive, predictive, and causal analysis provide a comprehensive understanding of behavior.

Descriptive analysis describes the data, like customer preferences and purchasing patterns, to see how many customers prefer to buy a product, for example. How are customers purchasing the items? If one buys a product A, then mostly the person will also buy a product T, and so forth.

Predictive analysis can be used to predict the future based on previous data. Most machine learning models belong to this type of analysis and help you to answer how many customers will buy product A and how many customers will buy product T.

Causal analysis will use cause and effect, the relationship between variables. This analysis helps you to determine the impact of specific behavior and to identify which variables are most influential in driving the behavior. Let's assume we have collected data on the price of product T and the corresponding number of customers who bought product A in the past. Here's an example of how we can conduct a causal analysis using this data:

1. Define the variables:

    • Independent variable: Price of product T (in this case, the price reduction to 50 percent).

    • Dependent variable: Number of customers buying product A.

2. Collect data: Let's say we have data for different price points of product T and the corresponding number of customers who purchased product A. Here's a hypothetical dataset:

    Price of product T (in % of original price) | Number of customers buying product A

    100% (original price) | 100 90% | 110 80% | 120 70% | 130 60% | 140 50% | ?? (unknown)

3. Analyze the relationship: We can now analyze the relationship between the price of product T and the number of customers buying product A. One way to do this is by using regression analysis, a statistical technique that examines the relationship between variables. Regression analysis can provide insights into how changes in the price of product T affect the number of customers buying product A.

4. Control for other factors: In the example, we're focusing on the impact of price reduction alone. However, it's important to consider and control for other relevant factors that might influence customer behavior, such as product quality, marketing efforts, competitor prices, or consumer preferences. This helps isolate the specific impact of the price reduction on customer behavior.

5. Draw conclusions: By analyzing the data and considering the statistical relationship, you can draw conclusions about the estimated impact of reducing the price of product T to 50 percent on the number of customers buying product A. In this case, the specific number of customers who will buy product A at a 50 percent price reduction cannot be determined without additional data or modeling. However, based on the observed trend in the data, you can make an estimate or prediction about the potential increase in customer purchases.

Data Collection

Data collection is an essential part of behavioral analysis because these data help you to gain insight into human behavior. I covered the data collection process in-depth in Chapter 4, "Datafication Pipeline."

Here are a few steps you should consider for behavior analysis:

- **Data Sources**: One data source is not enough to analyze the behavior of humans. Before choosing the type of sources, you need to define the objective and question for the behavior analysis. Based on this, you can choose the data sets and sources, like social media, hospital data, surveys, travel data, IoT data, etc.

- **Data Quality and Completeness**: The data set's quality is crucial for the outcome of your data analysis. It is essential to ensure that data is accurate, consistent, and recent.

- **Data Privacy and Regulations**: The data collection must comply with the privacy and local laws, such as GDPR in Europe, HIPPA in the USA, DPDPB in India, etc. This means you must obtain consent from the consumer for data collection and ensure data is stored securely.

- **Data Collection Method**: You can use various methods to collect data, such as real-time data with streaming and REST-based services and batch collection through SFTP. It is always good to have data collection through streaming.

- **Data Security**: Behavioral data is sensitive, so it is vital to have data security in place, like data encryption, data masking, data loss protection, data access, etc.

- **Data Storage**: Once you collect the data, you use data lakes or warehouses to store the cleansed data using cloud-native storage.

- **Training Data and Testing Data**: Ensure you have enough training data sets to train your model before using it.

Behavioral Science

Behavioral science is about human behavior and the factors that influence it. It draws from psychology, sociology, economics, and neuroscience and examines how a human thinks, feels, and behaves in different situations.

Behavior can be examined by involving mental processes that can be observed and studied. For example, when a human being is faced with a problem or situation, changing people's minds is not the same thing as influencing their actions; therefore, everyone engages in behaviors such as the following[1]:

- First, they analyze the situation by trying to understand the problem by breaking it down into smaller parts and examining each one.

- Second, they brainstorm with known people on a potential solution to address the problem.

- Third, if they have multiple options, they evaluate the pros and cons of each and weigh them against the goals.

- The evaluation allows them to make informed decisions based on the outcome of the step 4.

When you are collecting data or analyzing behavior, you need to consider the following:

- First is to define the research question and hypothesis using the data analysis framework I detailed in Chapter 5, "Data Analysis."

- Second is to select appropriate collection methods that can be used to collect the behavioral data, including surveys, observations, interviews, etc.

- Third is to adopt ethical standards by protecting privacy and confidentiality.

- Fourth is to analyze the collected data using the methods described in the following sections.

---

[1] *https://citationsy.com/styles/behavioral-sciences*

Importance of Behavioral Science

Behavioral science is essential in the present day for several reasons, as follows:

- **Mental Health**: Behavioral science can be used to identify and treat mental health disorders, such as anxiety, depression, and addiction. By understanding the underlying behaviors and thought patterns that contribute to disorders, behavioral science can help to create effective treatment plans.

- **Workplace Productivity**: By analyzing employee behavior and performance, behavioral science can help organizations to identify ways to improve workplace productivity and efficiency.

- **Healthy Lifestyle**: Behavioral science can encourage individuals to adopt healthy behaviors, such as a healthy diet, workout plan, etc.

- **Marketing Effectiveness**: Behavioral science can be used to understand consumer behavior and preferences, which helps the organization design marketing strategies.

How Behavioral Analysis and Analytics Are Processed

A behavioral analysis model is a method of studying human behavior to understand its underlying causes and motivations. It examines and analyzes specific behaviors of humans and summarizes human thoughts, feelings, and motivations.

Various models have their own methodology for conducting a study, and the following are a few analysis models.

Cognitive Theory and Analytics

Cognitive theory provides valuable insights into how our thoughts, beliefs, and cognitive processes influence our emotions, behaviors, and experiences. It highlights the active role of individuals in interpreting and making sense of the world around them, and it has had a significant impact on the field of psychology, particularly in understanding and treating various mental health conditions.

Cognitive theory uses a methodology to process and identify human behaviors. The following are a few steps:

- First, identify the problem that needs to be addressed. It could be any problem, like mental health, depression, etc.

- Second, collect the data by conducting an interview, facilitating a discussion, or doing self-reports.

- Third, use cognitive technologies such as natural language processing (NLP), machine learning (ML), and artificial intelligence (AI) to gain insight from complex unstructured data, such as text, speech, images, etc.

A few examples of cognitive theory and analytics are sentiment analytics, behavioral analytics, and image and speech recognition.

Biological Theories and Analytics

The biological model is used to study genetics, brain structure and function, and hormones to predict behavior. There are a few theories available to conduct research: genetic models, neurobiological models, endocrine models, and evolutionary models.

To conduct the analysis, you need to use biological and other types of data, such as demographic data, behavioral data, or environmental data. You then test hypotheses and identify patterns and relationships by using a model. They are as follows:

- **Regression Analysis**: It is used to identify a relationship between variables in data and is commonly used to identify the relationship between biological variables, like gene expression and hormone levels, and behaviors.

- **Structural Equation Modeling (SEM)**: This model tests the relationship between variables of biological factors, like genetics or brain structure, to see how they influence behavior.

- **Principal Component Analysis (PCA)**: This model is used to identify biological patterns from data and is used for brain imaging data.

- **Cluster Analysis**: This model groups biological data into clusters based on similarities. It is used in biological theory to group individuals based on biological factors, such as genetics and brain structure, and to identify how these groups differ in behavior.

- **Machine Learning**: This model is used in biological theory to identify patterns and relationships in large data sets and to predict future behavior based on data.

Integrative Model

This framework combines multiple theories to explain and predict behavior. It helps you to overcome the limitations of single theories by using multiple factors—such as biological, psychological, and social factors—and various analytics methods to identify patterns and relationships among the combined data.

The integrative model uses the following steps to process and analyze the data:

- First, collect the data from various sources using data integration and merging tools, including data lake, ETL, data mesh, data lineage, etc.

- Second, use statistical analysis models to identify patterns and relationships in the data. This can involve summary statistics, such as mean, median, and mode, and inferential statistics by using regression analysis.

- Third, use machine learning and NLP models to identify patterns and relationships and make a prediction based on those relationships using neural networks, decision trees, etc.

Behavioral Analysis Methods

Behavioral methods are used to identify, understand, and modify behavior patterns. The choice of method depends on the behavior being observed and the individual needs and abilities. The following are a few methods you can consider to identify behavior of humans.

Funnel Analysis

Funnel analysis is a method used to understand and optimize the user journey on any website. A funnel consists of a series of steps that a user goes through during the entire journey flow; for example, a shopping journey in an e-commerce application.

This method helps you to track user behavior at each step of the funnel and analyze the drop-off rates between each step to identify where the user stopped the process or exited the process. This will help the business to optimize the journey where the most users drop or exit the funnel.

For funnel analysis, you need to follow these steps:

- First, define the funnel's objective, like buying insurance, shopping cart journey, etc.

- Second, identify the goal of each step, like browsing the product catalog, entering payment information, etc.

- Third, collect data about the users and their completed steps, along with time spent on each stage, click-through rate, etc.

- Fourth, analyze the data using a decision tree, logistic regression, random forest, or neural networks. The choice of the algorithm depends on the objective of the analysis.

- Fifth, optimize the funnel based on the analysis and then redesign the funnel.

Cohort Analysis

Cohort analysis provides a valuable framework for understanding customer behavior and optimizing customer retention and segmentation strategies. It enables businesses to identify trends, make informed decisions, and ultimately improve customer satisfaction and business performance.

To perform the cohort analysis, you need to define the cohorts based on the relevant behavior—for example, behavior of the user's first purchase during the first week of every month—and define the cohort based on the first purchase. Once the cohort is defined, you can track the cohort against each metric, including retention rate, revenue per user, etc.

You can use regression analysis, k-means clustering, decision trees, survival analysis, and so forth, depending on the objectives of the cohort analysis.

Customer Lifetime Value (CLV)

The customer lifetime value (CLV) method is used to estimate the total amount of revenue that a business can expect to receive from a single customer over the course of its lifetime. This method takes input such as the customer's purchase history, the frequency and the value of their purchase, and the length of their loyalty.

```
CLV = (Average purchase value)*(Number of purchases per
year)*(Average customer lifespan)
```

The business can use the CLV value to make confident decisions on their business. If the CLV is more, then it is an excellent prospect to retain the customer. Various algorithms can be used to analyze CLV, including predictive analysis and historic CLV.

Churn Analysis

Churn analysis is the process of analyzing and predicting customer churn—when a customer stops using the product or service. This analysis examines customer behavior to identify the factors contributing to their decision to leave.

The churn analysis can be used to identify early warning signs, segment customers, predict churn, and analyze the impact of retention strategies.

You can use regression analysis, decision trees, random forest, neural networks, and survival analysis to analyze the data. These algorithms are based on the objective and data.

Behavioral Segmentation

Behavioral segmentation is dividing a group of customers into smaller groups based on behavior, action, and activities. Segmentation focuses on behavior rather than demographic or psychographic traits. While grouping, this process assumes customers behave similarly regarding their needs, preferences, and buying patterns.

There are several ways to approach behavioral segmentation, like purchase behavior, usage behavior, customer journey behavior, loyalty behavior, and engagement behavior.

Analyzing Behavioral Analysis

I mentioned in the previous section on the usage of prescriptive, descriptive, and causal analysis in behavioral analysis that all of these analyses are related to regression. Regression is the primary tool for all three analytics.

Descriptive Analysis with Regression

In these analytics, regression analysis summarizes and describes past data. Regression analysis is used to identify patterns and trends of past performance. The following is an example of how you can use R to perform a regression analysis for descriptive analytics for behavioral analysis.

Suppose we have a data set that includes the number of cakes and dessert sales per summer month and temperature. I use regression analysis to describe the relationship between each sale.

Note   I used modified data from Kaggle. It is available in GitHub.[2]

\# This data set is available in CC0:Public Domain[3] and I have modified the data set to add cake details along with ice cream to show the behavioral analysis.

```
cake_dessert_data <-read.csv("Cake_Dessert_Sales.csv")
summary(cake_dessert_data)
```
\# The output provides the min, median, and mean of cake_sales, dessert_sales, summer_months, and temperature.

```
      X              cake_sales         dessert_sales
summer_months       temps
 Min.    :1.0       Min.   : -3215    Min.     :-3290
Min.    :0.0000      Min.     :-3.287
 1st Qu.: 600.8     1st Qu.: 37089    1st Qu.:37565
1st Qu.:0.0000       1st Qu.:37.553
 Median :1200.5     Median : 54974    Median :55012
Median :0.0000     Median :55.039
 Mean    :1200.5     Mean    : 56698    Mean     :53451
Mean     :0.1667     Mean     :53.450
 3rd Qu.:1800.2     3rd Qu.: 76836    3rd Qu.:71178
3rd Qu.:0.0000     3rd Qu.:71.179
 Max.    :2400.0     Max.     :130637    Max.     :96508
Max.    :1.0000     Max.     :96.475
```

```
cor_matrix <- cor(cake_dessert_data)
corrplot(cor_matrix, method="square")
```

---

[2] https://github.com/shivugoniwada/Datafication
[3] https://www.kaggle.com/datasets/tysonpo/ice-cream-dataset

These commands are used to plot a circular correlation. Here, cor_matrix is a correlation matrix containing the pairwise correlations between variables. The method argument specifies the type of correlation plot. The method can be square, circular, etc.

The square correlation plot, as shown in Figure 7-1, is a graphical representation of the correlation matrix, with each variable represented by a point (line near 0.2) on the circumference of the square and the correlation between the variables represented by lines connecting the points. The strength of the correlation is indicated by the thickness and color of the lines, with thicker and darker lines indicating a stronger correlation.

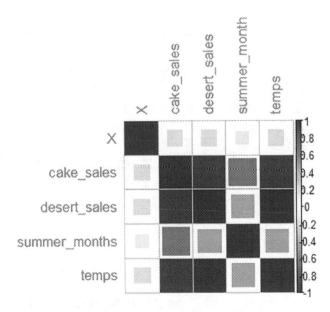

Figure 7-1. *Square correlation plot*

The resulting square correlation plot, as shown in Figure 7-1, provides a visual representation of the pairwise correlation between the variables in the data set. This helps you to identify patterns and relationships between the variables and gain insights into the behavior.

```
model <- lm(cake_sales ~ dessert_sales+summer_months, data =
cake_dessert_data)
summary(model)
```

The output is shown next. In this, you can find the following:

- Coefficient of the regression equation describes the relationship between dessert sales and summer months.

- The standard error of the coefficients—an estimate of the precision of the coefficients

- t-Statistics and p-value for each coefficient determine if the coefficients are statistically significant.

- R-squared value measures how well the model fits the data.

```
Residuals:
    Min       1Q   Median       3Q      Max
-25450.9  -3030.9     -8.4   2906.5  28557.5

Coefficients:
                   Estimate   Std. Error t value Pr(>|t|)
(Intercept)       22.215889   308.838382   0.072    0.943
dessert_sales      0.998668     0.005755 173.533   <2e-16 ***
summer_months  19775.646114   351.118593  56.322   <2e-16 ***
---
Signif. codes:  0 '***' 0.001 '**' 0.01 '*' 0.05 '.' 0.1 ' ' 1

Residual standard error: 5717 on 2397 degrees of freedom
Multiple R-squared:  0.9567,    Adjusted R-squared:  0.9567
F-statistic: 2.648e+04 on 2 and 2397 DF,  p-value: < 2.2e-16
```

```
# used to create a grid of four plots in a single graphics
window, as shown in Figure 7-2.
  par(mfrow=c(2,2))
```

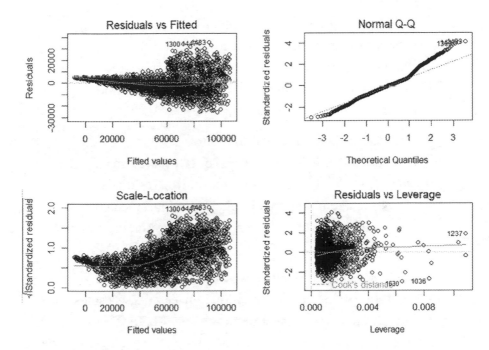

Figure 7-2. *Grid of four plots*

- **Residual vs. Fitted**: The plot shows the residual plotted against the fitted values. This plot helps you to identify any patterns or trends in the residual, which can indicate whether the linear regression model is a good fit for the cake dessert sales data.

- **Normal Q-Q (Quantile-Quantile**: This provides the dessert cake sales data following a normal distribution. In this, the quantiles of the observed data are plotted against the quantiles of the theoretical normal distribution. In the plot, they fall along a straight line. So, data is a normal distribution.

- **Scale Location**: A diagnostic plot that assesses the assumption of equal or same variance. The error variance is constant across all levels of the predictor variables. The points are randomly scattered around the horizontal line.

- **Residual vs. Leverage**: This plot provides the influential of individual observation. Observations with high leverage and large standardized residuals will be located in the upper right. This plot is located in the left-right, which means low leverage and highly standardized.

The plot of cake sales versus dessert sales in warmer months is shown in Figure 7-3. It shows a positive correlation between dessert sales and cake sales since both increase as temperature goes up. In the regression algorithm, cake sales using three variables at hand, the exploratory power of temperature on cake sales was added to the temperature variable, it means that the relationship between temperature and cake sales was investigated to determine its impact on the sales prediction, and it was noted that cake sales were forced to compensate for the temperature.

```
ggplot(data, aes(x=cake_sales, y=dessert_sales)) +
geom_point() +labs(x='cake_sales', y='dessert_sales') +
theme_classic()
```

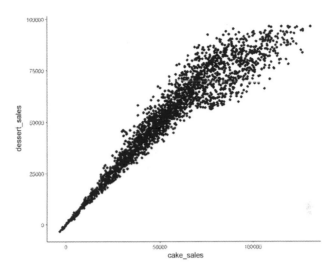

Figure 7-3. *Plot cake vs. dessert*

Causal Analysis with Regression

A causal analysis determines the cause-and-effect relationship between variables. By using causal analysis, you can identify the factors that influence behaviors. There are several ways you can use causal analysis for behavioral analytics.

The first approach is to use experimental design; in this, you can manipulate one or more variables to observe the effect of the behavior. For example, let's assign participants randomly to group 'A', and one group, either 'A' or 'B', is exposed to a particular stimulus or intervention and the other group is not. By comparing the behavior of the 'A' and 'B' groups, you can determine whether the stimulation or intervention caused a change in behavior.

The second approach is to use natural experiments, which involve the natural variations in an independent variable to identify its causal effects on a dependent variable. For example, you can compare the results of two people who live in different regions of the same city with varying levels of traffic to identify the effect of sleep on behavior.

The third approach uses observational data and statistical methods such as regression analysis to control confounding factors and identify causal relationships. For example, you may analyze data from a survey of ice cream and cold drinks consumption during warmer days. In this data, you can determine whether there is a causal relationship between a particular behavior and specific demographics.

Here is an example of world happiness index data for the year 2015 (Note: I have chosen only the first thirty countries).

In this example, I show how to conduct causal analytics for behavioral analysis using a regression model. I want to determine whether a causal relationship between country and happiness exists. Here's how you can use regression analysis to determine happiness.

```
# This data set is under CC0:Public Domain
data <-read.csv("HappinessData.csv")⁴
summary(data) – Provide the summary of data in my CSV file; in
this example, I am considering the entire data for analysis.
```

```
Country                Region              Happiness.Rank
Happiness_Score Standard.Error
 Length:29             Length:29           Min.   : 1
Min.   :6.575   Min.    :0.01848
 Class :character   Class :character   1st Qu.: 8
1st Qu.:6.853   1st Qu.:0.03411
 Mode  :character   Mode  :character   Median :15
Median :7.119   Median :0.03729
                                       Mean    :15
Mean    :7.096   Mean    :0.03919
                                       3rd Qu.:22
3rd Qu.:7.364   3rd Qu.:0.04176
```

---

⁴ *https://www.kaggle.com/datasets/mathurinache/world-happiness-report*

```
                                            Max.    :29
Max.   :7.587   Max.    :0.06476
 Economy..GDP.per.Capita.    Family         Health..Life.
Expectancy.    Freedom
 Min.   :0.9558              Min.   :0.9145   Min.   :0.6970
Min.   :0.4132
 1st Qu.:1.2502              1st Qu.:1.2196   1st
Qu.:0.8586             1st Qu.:0.5460
 Median :1.3263              Median :1.2802   Median
:0.8919            Median :0.6178
 Mean   :1.2951              Mean   :1.2428   Mean   :0.8743
Mean   :0.5894
 3rd Qu.:1.3601              3rd Qu.:1.3183   3rd
Qu.:0.9109            3rd Qu.:0.6404
 Max.   :1.6904              Max.   :1.4022   Max.   :1.0252
Max.   :0.6697
 Trust..Government.Corruption.   Generosity      Dystopia.
Residual
 Min.   :0.07785             Min.   :0.05841   Min.   :1.557
 1st Qu.:0.17521             1st Qu.:0.24434   1st Qu.:2.247
 Median :0.31814             Median :0.32573   Median :2.466
 Mean   :0.28642             Mean   :0.31403   Mean   :2.494
 3rd Qu.:0.38583             3rd Qu.:0.40105   3rd Qu.:2.676
 Max.   :0.52208             Max.   :0.51912   Max.   :3.602

happiness <- data$Happiness_Score
country <-data$Country
Check for linearity and normality assumption
```

#check for linearity as shown in Figure 7-4

```
ggplot(data, aes(x = country, y = happiness))+geom_
point()+geom_smooth(method = "lm")
```

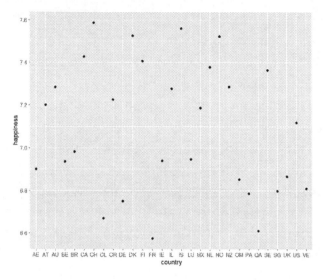

Figure 7-4. *Linearity plot*

The same data, when I compare economy against happiness, is as seen in the linearity graph in Figure 7-5.

```
happiness <- data$Happiness_Score
economy <-data$Economy
```

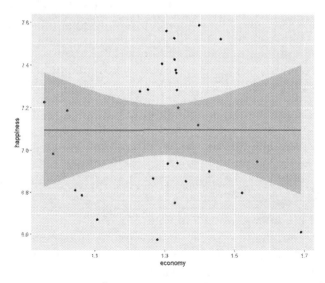

Figure 7-5. *Economy vs. happiness*

#Check for normality, the plot as shown in Figure 7-6

```
ggplot(data, aes(sample = residuals(lm(happiness ~ country,
data - data)))) +  stat_qq()
```

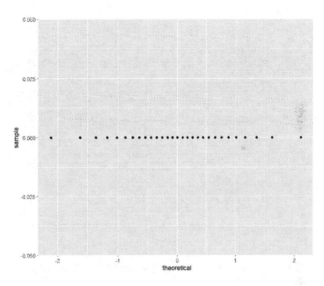

Figure 7-6. *Normality plot*

#Check for normality for economics plot as shown in Figure 7-7

```
ggplot(data, aes(sample = residuals(lm(happiness ~ economy,
data = data)))) +  stat_qq()
```

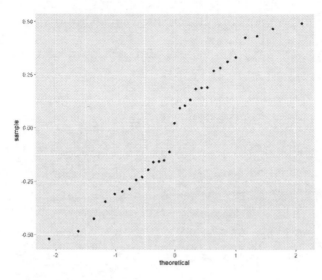

Figure 7-7. *Normality for economies plot*

Run regression analysis

```
reg_model <- lm(happiness ~ country, data = data)
```

#check the regression summary

```
Summary(reg_model) - following is the regression summary:
```

Residuals:
ALL 29 residuals are 0: no residual degrees of freedom!

Coefficients:

	Estimate Std.	Error	t value	Pr(>|t|)
(Intercept)	6.901	NaN	NaN	NaN
countryAT	0.299	NaN	NaN	NaN
countryAU	0.383	NaN	NaN	NaN
countryBE	0.036	NaN	NaN	NaN
countryBR	0.082	NaN	NaN	NaN
countryCA	0.526	NaN	NaN	NaN
countryCH	0.686	NaN	NaN	NaN

countryCL	-0.231	NaN	NaN	NaN
countryCR	0.325	NaN	NaN	NaN
countryDE	-0.151	NaN	NaN	NaN
countryDK	0.626	NaN	NaN	NaN
countryFI	0.505	NaN	NaN	NaN
countryFR	-0.326	NaN	NaN	NaN
countryIE	0.039	NaN	NaN	NaN
countryIL	0.377	NaN	NaN	NaN
countryIS	0.660	NaN	NaN	NaN
countryLU	0.045	NaN	NaN	NaN
countryMX	0.286	NaN	NaN	NaN
countryNL	0.477	NaN	NaN	NaN
countryNO	0.621	NaN	NaN	NaN
countryNZ	0.385	NaN	NaN	NaN
countryOM	-0.048	NaN	NaN	NaN
countryPA	-0.115	NaN	NaN	NaN
countryQA	-0.290	NaN	NaN	NaN
countrySE	0.463	NaN	NaN	NaN
countrySG	-0.103	NaN	NaN	NaN
countryUK	-0.034	NaN	NaN	NaN
countryUS	0.218	NaN	NaN	NaN
countryVE	-0.091	NaN	NaN	NaN

Residual standard error: NaN on 0 degrees of freedom
Multiple R-squared: 1,   Adjusted R-squared: NaN
F-statistic: NaN on 28 and 0 DF, p-value: NA

```
# run regression analysis for economy

reg_model <- lm(happiness ~ economy, data = data)
summary(reg_model)
Residuals:
     Min       1Q   Median       3Q      Max
-0.52078 -0.24301  0.02289  0.26807  0.49089
```

```
Coefficients:
            Estimate Std. Error t value Pr(>|t|)
(Intercept)   7.0922      0.4536  15.637 4.69e-15 ***
economy       0.0028      0.3474   0.008   0.994
---
Signif. codes:  0 '***' 0.001 '**' 0.01 '*' 0.05 '.' 0.1 ' ' 1

Residual standard error: 0.3122 on 27 degrees of freedom
Multiple R-squared:  2.407e-06,   Adjusted R-squared:  -0.03703
F-statistic: 6.499e-05 on 1 and 27 DF,  p-value: 0.9936
```

The regression analysis will produce a summary that includes the estimated coefficients for the country and economy, the standard error of the coefficients, the t-value, and the p-value. The p-value indicates whether the effects of country and economy on happiness are statistically significant. For example, if the p-value is less than 0.05, you can conclude that there is a statistically significant relationship between economy and happiness. A positive coefficient would indicate that the economy is positively related the happiness, while a negative coefficient would indicate that the economy is negatively related to happiness. In the preceding economy example, the coefficients are positive and p-values are less than 0.05.

Causal Analysis with Experimental Design

For causal effect calculations, you must use another method, such as experimental design. Here is an example of mediation. I will provide details on whether a mediation program can increase happiness.

Let's use the data of the happiness treatment group, which is the list of people who participated in mediation, and the other group, the happiness control group, which is the list of people not participating in the mediation.

The following program measures the data before and after mediation and compares the change in happiness levels between the two groups.

```
library(randomizr)
set.seed(123) # for reproducibility
allocation <- complete_ra(N = 50, m = 25)
#This creates a vector allocation with values 0 and 1, where
0 indicates the happiness control group and 1 indicates the
happiness mediation group
```

```
#The data collected on participant's happiness levels before
and after the mediation, I simulated some data
```

```
set.seed(123) # for reproducibility
```

```
#before_mediation is a vector of pre-mediation happiness level,
and after_mediation is a vector of post-mediation levels. The
'if else' assigns a different mean and standard deviation to
the mediation and control groups.
before_mediation <- rnorm(50, mean = 50, sd = 10)
after_mediation <- ifelse(allocation == 1, rnorm(25, mean = 60,
sd = 8), rnorm(25, mean = 50, sd = 8))
```

```
#Finally, t.test to compare the change in happiness between the
two groups. This calculates a t-test with a mean difference
between the two groups as the test is static. The output will
give the p-value and confidence interval for differences in
means, with which you can assess the statistical significance
of the mediation effect.
t.test(after_mediation - before_mediation ~ allocation)
```

```
#the result is as follows.
```

```
Two Sample t-test
```

```
data:  after_mediation - before_mediation by allocation
t = -3.6406, df = 47.826, p-value = 0.0006668
alternative hypothesis: true difference in means between group
0 and group 1 is not equal to 0
95 per cent confidence interval:
 -17.982208  -5.185697
sample estimates:
mean in group 0 mean in group 1
       2.128706        13.712658
```

Here the p-value is 0.0006668, which is less than 0.05, so there is evidence to support the hypothesis that mediation can increase happiness.

Challenges and Limitations of Behavioral Analysis

As I explained in previous sections, behavioral analysis is used in various fields, such as psychology, criminology, biology, and marketing, to gain insight into human behavior. However, it has its limitations and challenges, such as the following:

- **Ethical Concerns**: Behavioral analysis involves invasive methods that concern ethics, such as tracking human activity, behavior, etc.

- **Data Samples**: Behavioral analysis requires a huge chunk of data. Sometimes, you may not get the necessary data, so you may not be able to analyze with this small sample size.

- **Predictability**: The behavioral analysis can identify patterns of behavior, but it cannot predict future behavior with clarity as human behavior is influenced by various factors. Getting data on multiple aspects is challenging.

- **Data Interpretation**: The interpretation of data in behavioral analysis is subjective and can be influenced by personal biases and can lead to conflicting results.

Summary

Behavioral analysis is used to understand and predict human behavior in the fields of marketing, business, psychology, and biological science. In the field of marketing and business, behavioral analysis will predict customer purchasing behavior by analyzing how customers interact with their product or services and can help identify patterns and trends. The fields of psychology and biology help you to identify human and animal behavior and identify patterns and trends, as well as influencing factors. Also, it uses ethe cognitive processes, such as learning, memory, decision making, social behavior, and more.

In this chapter, I tried to provide an introduction to behavioral analytics, as behavioral analytics is a vast topic. You can refer to various books that specifically examine behavioral analytics.

CHAPTER 8

Datafication Engineering

Datafication engineering is designing, developing, and deploying artificial intelligence (AI) and machine learning (ML) systems. It involves a range of tasks, including data collection and preparation, model selection, deployment, hyperparameter tuning, model training and evaluation, and deployment and monitoring of the models.

In this chapter, I am covering all the details mentioned here, including the automated process of data ingestion and data processing.

This chapter will answer your crucial engineering questions.

Steps of AI and ML Engineering

AI and ML engineering is an essential part of any AI and ML project, and it involves model design, development, training, testing, and deployment into the production system. AI and ML are engineered through various approaches and techniques depending on the specific problem and objective.

© Shivakumar R. Goniwada 2023
S. R. Goniwada, *Introduction to Datafication*,
https://doi.org/10.1007/978-1-4842-9496-3_8

Following are the steps you need to follow when developing new AI and ML projects:

- **Problem Definition**: This is an essential step of any AI and ML project. In this step, you need to define the problem that you are trying to address. Based on the problem, you must prepare a questionnaire and define the input and output type using the data analysis framework explained in Chapter 5, "Data Analysis."

- **Identify Source**: This is one of the most vital steps in the entire engineering process because the data type is crucial for any output. You must identify the source based on the questionnaire you did in the first step. For example, if you want to analyze cancer-related diagnoses, then the data source is various cancer hospitals.

- **Data Collection**: This step is to collect the data from the source through streaming, REST services, or batch jobs, depending on the source technologies.

- **Data Processing**: This step is to cleanse and format the data, and then select relevant features and split the data into training, validation, and test data sets.

- **Model Selection**: You select an appropriate model based on the questionnaire and data sets. The model can be a decision tree, neural network, and so on.

- **Statistical Prediction Modeling**: In this step, you develop a code for the model you selected using programming languages like Python, R, etc.

- **Train the Model**: In this step, you must train the model you developed using the training data. This involves adjusting the model parameters to minimize error or loss of function.

- **Interpretation of the Result**: So far, you have done the development and analysis and calculated some results. In this step, you need to interpret the findings using the evaluation and validation data. This involves comparing the predicted outputs to actual outputs and calculating metrics such as accuracy, precision, and recall.

- **Challenging the Result**: In this step, the model parameters are further fine-tuned and optimized by performing hyperparameter tuning using settings such as learning rate, regulation, etc.

- **Deployment**: Once the model is trained and validated, you need to deploy it to make predictions on the new live data set.

AI and ML Development

It is the process of creating, implementing, and refining AI and ML models and systems. it involves various stages and techniques to design, train, evaluate, and deploy intelligent systems that can learn from data and perform tasks without explicit programming.

Understanding the Problem to Be Solved

This step is critical in AI and ML development. In this, you identify the objective or problem and the outcome you are looking for. Based on the type of question, you will identify data sets and models.

The type of question can be descriptive, exploratory, inferential, predictive, causal, or mechanistic, as described in Chapter 5, "Data Analysis."

217

Based on the type of question, you need to define the success criteria by identifying the metrics that will be used to evaluate the model's performance. Understanding the constraints and risks is an essential step in any AI and ML development.

Choosing the Appropriate Model

Various models exist for AI and ML engineering. You can use a single model or a combination of many models to arrive at the desired results. Choosing the suitable model involves understanding the nature of the question and the problem to be solved. There are many models available, as follows:

- **Statistical Models**: The statistical models are used to describe the relationship between different variables and to make a prediction based on data. They can be divided into descriptive and inferential models. Statistical models are used to analyze data and make predictions based on statistical theory. The models include linear regression, logistic regression, and time series analysis.

- **Machine Learning Models**: ML models are used to perform predictive analysis on data and to make predictions based on that learning. These models are designed to automatically improve performance by learning from past data. There are three main types of ML model: supervised learning, unsupervised learning, and reinforcement learning. Supervised learning training is based on a labeled data set, and unsupervised learning is based on an unlabeled data set. The essential and most common models are linear regression, logistic regression, neural networks, decision trees, support vector machines, random forests, and deep learning.

- **Simulation Models**: Models are used to simulate real-world situations to help understand and predict the behavior of a system. These models are helpful in digital twin that is virtual representation of a physical object, process or system and Internet of Things (IoT) analysis. The most common models are agent-based models, Monte Carlo simulation, and discrete event simulation.

- **Cognitive Models**: Models are used to analyze the behavior of the human brain to discover how people think, learn, and make decisions. These models are based on the idea that human cognition involves the manipulation of the mental representation of information, which is processed and transformed through a different cognitive process. These models are commonly used in neuroscience, psychological research, and other AI research. The most common cognitive models are deep learning, neural network, natural language processing (NLP), and expert systems.

Choosing a suitable model is an essential step in AI and ML engineering. Here are a few guidelines you may need to follow when selecting the appropriate model for your analysis:

- Before choosing any model, you need to clearly understand the problem and the objective, as described in the previous section.

- It is essential to understand the data sets, including data distribution, data sources, data variables, and metadata.

- It is essential to understand the assumption in each model. For example, linear regression assumes that the relationship between the independent and dependent variables is linear.

219

- Always keep performance in mind, so evaluate the model performance on the selected data set by using various methods such as accuracy, precision, recall, and F1 score. If your selected model is not up to the performance you expected, try to refine the model by changing the hyperparameters by modifying the features, or choose a different analysis model.

Preparing and Cleaning Data

Preparing cleansed data is crucial for any AI and ML project. As I explained in Chapter 4, "Datafication Pipeline," there are various methods you can use to get cleansed data from multiple sources. The model provides output based on what is learned. Feeding the model with low-quality data won't produce results as expected, regardless of how good your model is. While getting data from sources, you need to have a few guidelines in place, as follows:

- Ensure the data is labeled correctly and consistently. Labeling errors significantly impact the performance of your model. The data is processed by adding one or more tags so that the model can learn from that.

- A few companies have been collecting data for years for AI and machine learning. You can use these data for your analysis. Almost all mature governments provide open data, and online data sets are available. They are Kaggle, google fusion tables, CKAN, Quandl, and data market.

- Check your data for missing values or duplicates and take necessary action to enrich or remove the data. It is best to spend some quality time analyzing the data's accuracy.

- Remove the errors in the data. Erroneous data impact the performance of your model.

- Use augmentation techniques, including latent
  semantics and entity augmentation, if you don't have a
  big enough sample data set.

- Validate and test the data set by checking for
  consistency, accuracy, and completeness. Test your
  data by splitting testing and training sets to ensure that
  your AI and ML models can generate new data.

Overall, getting cleansed data for AI and ML requires a combination of careful planning, data collection, and data cleansing processes to ensure that your data is reliable, accurate, and relevant to your needs.

Feature Selection and Engineering

Once you have the model and data set identified, the next step is to select a helpful feature or variables from the data set to improve performance and reduce your model's overfitting and computational complexity.

Assume you have a data set with millions of rows. This data set may contain unwanted features irrelevant to your analysis. The main goal is to identify a subset of features that includes the most relevant information while reducing and minimizing the redundant information.

There are many techniques available to identify the features from your overall data set, including correlation-based feature selection, recursive elimination, and principal component analysis.

Correlation-Based Feature Selection

The correlation-based feature analysis is the most common and popular technique used by most AI and ML projects. It uses the correlation matrix between the features or variables and the target variable using methods like the Pearson's correlation coefficient or Spearman's rank correlation coefficient. After the correlation matrix, the next step is to rank the feature based on the matrix and select the top-ranked features for your model.

Recursive Elimination Feature Selection

Recursive elimination uses automation to identify the data set's features without having prior knowledge of the problem. It uses an iterative process to recursively remove features from the data set and train the model with the remaining feature set until the optimal feature is reached. From the optimal feature, again, it ranks based on its importance and contribution to the model. You can use various methods, such as linear model coefficients or the decision tree's feature importance.

Principal Component Analysis (PCA)

The PCA reduces the number of features in the data set while retaining the essential information. This occurs by identifying the directions of maximum variance in the data set and projecting data onto a new coordinate system defined by these directions.

Model Training and Optimization

Model training is the most essential step in overall AI and ML engineering. The training depends on the type of question, the model you select, and the data set. There are several different ways to train the model as follows:

- **Supervised Learning**: The model is trained with labeled data, where each example has an input and an output. The objective is to learn a model that maps the input and output.

- **Unsupervised Learning**: The models are trained on unlabeled data, where there is no predefined output. The objective is to discover the patterns and structure in the data set.

- **Semi-supervised Learning**: The models are trained
  with labeled and unlabeled data. The objective is to
  leverage the unlabeled data to improve the model's
  accuracy.

- **Reinforcement Learning**: The model learns by
  interacting with an environment and receiving
  feedback through rewards or penalties.

- **Transfer Learning**: The model is pre-trained on a
  large data set and then fine-tuned on a smaller data set
  related to the original task.

AI and ML Testing

AI and ML testing is a crucial component of the entire development
process, and the aim is to ensure that the system behaves as expected. As
per testing, there is no difference in the testing lifecycle of web application
testing and AI and ML testing.

Both developments undergo unit testing, integration testing,
regression testing, non-functional testing, security testing, and exploratory
testing (fitness testing).

Unit Testing

Unit testing involves testing individual modules or algorithms to ensure
they are working as expected. To perform unit testing, you need to break
complex models or algorithms into smaller units, and each unit is to be
tested with the test cases.

The test data is a vital component in the entire testing process, using
the validation data set that covers a wide range of possible input and
output values and scenarios with unusual patterns.

Example 1

Assume that your AI system uses machine learning to identify images of an eagle and sparrow. Your AI system is based on the neural network model that takes an image as input and output to predict whether the image contains an eagle or a sparrow.

To conduct unit testing for this example, you need to identify the individual units of the entire AI system that can be tested in isolation. For this example, a neural network would be a suitable unit for testing.

For test data, you could create test images of eagles and sparrows with different styles, backgrounds, colors, along with bird images other than eagles and sparrows. These data allow you to test with various negative and positive scenarios. For the actual scenarios, use mock data to test the model in isolation with synthetic images of eagles and sparrows using image generation libraries available on the internet.

During the entire unit testing, monitor the behavior of your model and analyze the test result to identify issues and errors.

Use unit testing frameworks such as PyTest and libraries like NumPy and Tensor Flow to create unit test cases. To automate unit testing, integrate PyTest into the DevOps Pipeline with Jenkins.

Example 2

Assume that your AI system uses the linear regression model to predict the price of the camera based on the lens, image processing, brand, pixel, etc. The entire model was developed with linear regression by using the R program or Python.

To conduct unit testing for this example, you need to identify the individual units of the entire AI system that can be tested in isolation. For example, a linear regression model would be suitable for unit testing. The next step is to develop test cases with test data that cover a range of possible input values and scenarios, like different lens size, model type, lens type, number of pixels etc.

If your code is developed using R, then use "test that" or "tiny test" - in tiny test, tests are scripts, interspersed with statements that perform checks to write test cases and use PyTest for Python and automation. You could integrate these test cases into a DevOps pipeline, like Jenkins.

Integration Testing

In unit testing, you isolate every component of your model. In integration testing, you combine all your model components to ensure it all works together to achieve the desired outcome.

Example:

Let's take the previous unit testing example of camera defects. As I stated in unit testing, your model involves multiple components, such as lens, capturing images, and image processing, to detect and estimate the price of the camera and connects to Google drive or other storage through the internet.

Next is to create integration test cases that cover the range of possible scenarios, pixels, behavior in various lightning conditions, etc. and deploy all the components into the cloud to mimic real scenarios to test the end-to-end process to identify camera defects.

Monitor and analyze the output generated by the test cases to certify whether the camera is in good condition or has some errors.

To automate integration testing, integrate test cases with the DevOps pipeline with Jenkins or another CI orchestrator.

Non-Functional Testing

Non-functional testing is about testing your AI and ML model to ensure it will perform as expected in the production environment. The following are the few non-functional "ilities." A few of these are to be taken care of during design and development, and a few are to be managed during runtime.

- **Performance**: The ability of the model to perform a data size (large or low) of functional model or transaction

- **Compatibility**: The ability of your AI and ML system to exchange data and integrate with other components of your overall ecosystem

- **Recoverability & Availability**: The ability of your AI and ML system to recover from the failure of one component and be available as required

- **Reusability**: The ability of your AI and ML sub-components to be reused across your organization in other AI and ML projects. Define a library of components and provide open integration protocols to integrate with the other components of the AI system to increase reusability.

- **Operability**: The ability of your AI and ML components to be designed for operations, like ease to change, modification, and adding new features.

- **Security**: A critical NFR "ility'" for AI and ML systems. It helps you to secure data and access from encryption, masking, and authorization perspectives.

- **Traceability & Auditability**: The ability of the AI and ML system to ensure the accuracy and consistency of the data and process over the entire life cycle of the model

- **Scalability and Elasticity**: The ability of the AI and ML system to handle data size and potential to be scaled to accommodate that growth by using cloud platform.

Performance

Performance testing is one of the most critical NFR tests because you must ensure that your model can handle large amounts of data, process it efficiently, and provide an accurate and reliable result. The size of the data set can be terabytes or gigabytes of data. While doing performance testing, you need to consider a few things, such as the following:

- **Data Size**: As I mentioned, the size of the data set used to train your model can significantly impact its performance. You must consider all sizes of data set from small to large to evaluate the behavior of your model on different sizes.

- **Latency**: The latency is another key performance metric you must consider. Performance testing should measure the latency of your system and check which component in the model has the issue.

- **High Availability and Scalability**: In an AI and ML system, the data is not fixed, so testing must include testing the scale to meet the data size. For example, social media data is not constant, and the data size may increase during holidays or evenings rather than during business hours.

To measure the preceding metrics, your system should conduct various tests under the performance umbrella, as follows:

- **Benchmark Testing**: Benchmark testing involves testing the performance of your AI and ML system against the industry benchmark data set. There are benchmarks available like ImageNet, which is used to test the performance of image recognition systems.

- **Load Testing**: Load testing is about testing your system against the data and users. This testing helps you to identify how your model behaves during abnormal situations beyond capacity.

- **Stress Testing**: Stress testing involves testing the model in an abnormal condition. With this test, you can check the behavior of your system against the memory, CPU, etc.

- **Fitness Testing**: Fitness testing involves testing your model with real-world scenarios to identify how it performs in real-world functionality and non-functionality perspectives. This type of testing uses A/B testing to simulate the real users.

Various tools and frameworks are available to test the performance of your AI and ML systems. Tensor Flow Performance Dashboard is used for Tensor models; MLPerf is open source and tests the performance of the ML models; Keras Tuner is a hypermeter optimization framework that solves issues of the hypermeter search; ML Flow helps you track things such as metrics, parameters, and artefacts across multiple runs of a model of the ML experiments; and finally, Apache JMeter helps you to do load and stress testing.

Security Testing

Security testing is crucial for AI and ML systems because these systems are vulnerable to several attacks, such as data poisoning, model stalling, model corruption, and adverse attacks. Besides model security, the data must be protected because of its sensitivity and personal information. Besides security threats, you must test against various country-specific compliance regulations.

To make your system more secure and stable, you need to conduct various security testing, such as adversarial testing, data poisoning testing, privacy testing, robustness testing, and accessibility testing. Along with these application-level security tests, you need to carry out penetration testing, vulnerability testing, and so on.

- Adversarial testing helps you to test your system's robustness against adversarial attacks, such as maliciously crafted inputs that can cause the system to behave in unexpected ways.

- The data poisoning testing helps you to test against data injection against the training data set to manipulate the system behavior.

- Privacy testing helps you to test the user's privacy and manipulation of sensitive data.

- Access testing helps you to test the user access and permission against the data at rest and data in motion for training an actual data set.

- Penetration testing helps you to test the perimeter security vulnerability, such as firewalls, keys, secrets, and port- and subnet-level testing. This testing allows you to identify weakness in the model's security defense.

There are various tools and frameworks that can be used to test the models. They are as follows:

- **Threat Modeling:** This provides a model that identifies potential threats to the AI and ML model and develops strategies to mitigate those threats. Threat modeling is one of the critical frameworks to identify areas of weakness in the model's security defense and plan to address those weaknesses. I suggest you create a threat model for all your AI and ML systems. You can use tools like Microsoft Threat Modeling to design threat modeling.

- **Code Review**: This involves reviewing the code to identify security vulnerabilities and technical debt using static and dynamic code analysis tools. There are various open-source and vendor-specific tools available, such as Microfocus Fortify, Nessus, Qualys Guard, Burp Suite, etc.

- **Privacy Testing**: You can use tools like TensorFlow Privacy to test the privacy of individuals in the data used for training.

- **Adversarial Testing**: You can use the Adversarial Robustness Toolbox and CleverHans, which can provide tools for adversarial testing and testing the robustness of your models.

DataOps

"DataOps is a collaborative data management practice focused on improving communication, integration, and automation of data flows between data managers and consumers across an organization."

— Gartner

"The Goal of DataOps is to create predictable delivery and change management of data, AI and ML models, and related artefacts. DataOps seeks to reduce the end-to-end cycle time of data analytics, from the origin of ideas to the literal creation of charts, graphs, and models that create values."

— Gartner

DevOps and DataOps share many similarities, with a few specific capabilities used in software engineering and data engineering.

DevOps emphasizes collaboration, automation, and continuous improvement for software development products. DevOps include practices such as continuous integration, continuous delivery, continuous deployment, infrastructure as code, and configuration as code. The DevOps workflows are more code-centric and code quality.

On the other hand, DataOps emphasizes collaboration, automation, and continuous improvement and managing and delivering datafication products. The DataOps workflows are more data-centric, data quality, model training, and model deployment and monitoring. DataOps focuses on ensuring data is accurate, relevant, and free from biases before it is used to train the model.

DataOps is a set of practices and principles that aim to improve the speed, efficiency, and quality of data-driven decisions. DataOps is based on the principles of DevOps and Agile methodologies, and it focuses on collaboration, automation, and continuous improvement.

DataOps involves the entire life cycle of data, from data preparation and ingestion through data processing, analysis, and visualization, as shown in Figure 8-1, to the deployment and operation of the datafication process. The DataOps emphasizes various teams, including data engineers, data scientists, business users, data architects, and enterprise architects, to ensure datafication is delivered quickly and in an automated fashion.

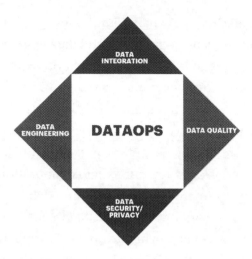

Figure 8-1. *DataOps capabilities*

The key best practices and principles in DataOps are as follows:

- **Automation**: Automation is a vital part of DataOps.
  It allows teams to streamline the data workflows and
  reduce errors. By automating repetitive tasks, you can
  improve the overall quality with reduced effort.

- **Continuous Integration and Delivery (CI/CD)**:
  DataOps uses CI and CD to ensure the datafication is
  tested and deployed quickly and reliably.

- **Agile Methodologies**: DataOps can use Agile
  methodologies to ensure that datafication is developed
  iteratively with an integrated backlog and regular
  feedback.

- **Data Governance**: DataOps ensures that data is
  accurate, reliable, and compliant with relevant
  regulations and standards.

- **Observability**: DataOps can integrate with observability tools to monitor and alert about any detected issues in datafication products in near real-time.

By using DataOps, you can have several benefits, such as faster time to market, quality improvements, increased collaborations, better decision making, ease of operation, and continuous improvement.[1]

MLOps

The AI and ML systems require thoughtful management of the AI life cycle and operational due diligence to ensure the objectives of the AI and ML systems are achieved. It is the process of developing an experimental model into a streamlined testing and production environment.

MLOps uses a set of standardized processes and technology capabilities to operationalize and scale the AI and ML development life cycle. It uses DevOps practices for development, deployment, and management. It involves the integration of ML workflows into the software development cycle.

The MLOps life cycle, as shown in Figure 8-2, involves nine steps, each with its own process.

---

[1] Cloud Native Architecture and Design Patterns, *Shivakumar Goniwada, Apress, 2021*

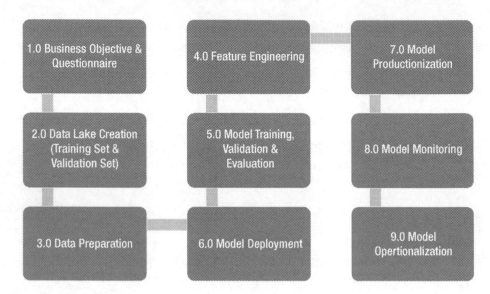

Figure 8-2. *MLOps life cycle*

Processes 1 and 2 involve business alignment and data readiness, and processes 3, 4, and 5 are involved in building the model, which consists of model coding, data processing, training, inferences, test automation, and security and compliance check. Processes 6 and 7 are involved in deployment and execution. Processes 8 and 9 are to be managed in post-production to carry out monitoring of the models and operation of the model.

The technical capabilities required to support AI and ML model development and experimentation are as follows:

- Development workplace where engineers can access required data and tools to build and test AI models and where modeling tools are tools for authoring custom training logic.

- Feature engineering is engineering the right features for the model from available data, either labeled or unlabeled. You can use auto feature engineering that automatically decides model performance feedback to re-select or engineers the feature that yields the best results.

- Feature store is used to store features in repositories for updating, retrieving, sharing, and discovering curated ML features for reuse across different models. You can also store metadata of the features, including definitions and data sources.

- Model training to train the model from input data, algorithms, code, and hyperparameters.

- Experiment management is to manage the ML experiments to be able to track or replicate them to improve, debug, or reproduce them. This includes tracking the hyperparameters, algorithms, code versions, data sets, logs, model experiment results, and resource metrics.

Summary

Datafication engineering is a specialized process within the overall software engineering process that focuses on development and deployment of datafication models into the production environment. As explained in the chapter, the main key practices of engineering are data preparation, model development, model training, model evaluation, and model and data deployment.

Datafication is a rapidly evolving area with new technologies, tools, and techniques being developed, and this requires a combination of processes, skills, and methodologies to automate the process.

Successful engineering requires a collaboration across various stakeholders, such as the business, enterprise architecture, data architecture, data scientist, data engineers, model engineers, and quality engineers.

In this chapter, I have provided an introduction to the engineering process, pipelines, and collaboration.

CHAPTER 9

Datafication Governance

The rapid development of datafication technologies has transformed various aspects of our lives across almost all industries. However, these technologies also bring many challenges that require careful consideration and execution of governance.

The governance of datafication involves development frameworks that ensure that these technologies are developed and used ethically, transparently, and accountably while minimizing risks to individuals.

Governance requires a multi-disciplinary approach that involves data architects, data engineers, software engineers, data scientists, machine learning (ML) engineers, chief data officers (CDO), chief privacy officers (CPO), society at large, legal experts, and policy makers.

This chapter reviews the governance framework, along with risk management processes, principles, and an introduction to regulations. By addressing the governance challenges associated with artificial intelligence (AI) and ML, you can ensure that these technologies are developed and used to benefit society and respect human rights and dignity.

© Shivakumar R. Goniwada 2023
S. R. Goniwada, *Introduction to Datafication*,
https://doi.org/10.1007/978-1-4842-9496-3_9

Importance of Datafication Governance

Datafication is a rapidly advancing field with the potential to significantly transform organizations. However, as with any new technology, its management has risks and challenges. Datafication governance is essential for accountability, eliminating bias, compliance, privacy, security, quality, and acceptance.

Datafication governance is to be defined as the management of the data and AI/ML model assets. Governance aims to ensure that the data and models are appropriately managed according to policies and best practices. The governance focuses on how decisions are made about data and how AI/ML models are stored, processed, trained, and managed.

Datafication governance must be done in collaboration. It needs to align directly with the organization's strategy. With this alignment, people in the organization and industries will likely change their behavior and adopt governance practices.

A lack of robust governance impedes the success of your datafication program, and it could put your business in peril. According to the survey conducted by Algorithmia's "2021 enterprise trends in ML"[1] report, around 58 percent of organizations struggle with governance, security, and audibility. This shows that governance is the top challenge in the datafication area.

Datafication governance is not a one-time activity. It requires a constant focus on ensuring that an organization gets value from its data and AI/ML models and reduces the associated risks. The datafication governance team is a virtual organization with specific accountabilities headed by a chief data officer (CDO).

---

[1] *https://aithority.com/machine-learning/algorithmia-report-reveals-top-enterprise-trends-in-artificial-intelligence-machine-learning-for-2021/*

Why Is Datafication Governance Required?

The datafication governance is required to ensure responsible and ethical use of data in the era of increasing data collection, processing, and analysis. All industries, primarily financial, health care, and social, are heavily regulated by evolving legislation from the government. Governance helps you to ensure compliance with these regulations, reducing the risk of legal and reputational damage.

Datafication governance is more often focused on reducing the risks, which has the potential to impact a wide range of business processes, including strategic decision making, customer service, product development, and sales inclusion. The governance framework can help identify and mitigate potential risks associated with development and deployment of AI and ML models.

Another important factor related to the development and deployment of AI/ML models is that the result during this process may vary depending on the data set. The result during training may differ from that during production. Take care of training and production biases in which governance plays a pivotal role. There are two sides to any model. The wrong side of the model emerges with biases and other issues; for example, the insurance company whose models recommend offering to lower auto insurance to senior citizens compared to adults in the same city.

Datafication Governance Framework

Datafication governance is challenging, requiring multiple organizations and stakeholders to come together to drive the governance. As shown in Figure 9-1, governance for AI/ML models is critical to address regulation issues, risks, and other challenges. The framework must ensure that data scientists and researchers follow governance framework when building the models.

Figure 9-1. *Datafication governance framework*

Oversight and Accountability

The data and models must adhere to policy and compliance requirements established by each industry sector or government, and respective stakeholders within the organization, broader society, and other government institutions must review and approve each stage of the model lifecycle.

The control and testing process must ensure the reliability, performance, and safety of the models. This process helps you to identify potential risks and issues and validate the effectiveness of the models. The performance metrics must be designed to gain assurance that the model is meeting expectations. The choice of the metrics depends on the problem type, like classification, regression, clustering, and the type of questionnaire of AI/ML models. A few performance metrics are accuracy, precision, recall, F1-sore, mean absolute error, mean squared error, R-squared, adjusted Rand index, and mean average precision.

To control the model, various tools are available for respective areas, such as Pandas, Scikit-learn for data preprocessing and feature selection, Tensor Flow, and PyTorch for model development and validation. There is also Optuna, Hyperopt for hyperparameter optimization, MLFlow, TFX for model monitoring and management, SHAP, LIME for model explainability and interpretability, AI Fairness 360, and Fairlearn for fairness and bias detection.

Model Risk, Risk Assessment, and Regulatory Guidance

Datafication risk refers to the potential for negative consequences from the models. These risks arise due to biases, data quality issues, and overfitting. Performing a risk assessment can help you to identify and quantify risks and take necessary mitigation steps. Although AI/ML brings many benefits in time, they can pose novel challenges for organizations and regulators and amplify existing risks to consumers.

To mitigate these risks, you can take the following steps:

- Conduct a risk assessment.

- Find strategies to mitigate identified risks.

- Have transparency in AI/ML models.

- Enact regulatory monitoring.

- Validate and train models.

- Communicate with stakeholders.

Various countries define regulations and guidelines: SR11-7, OSFI E23, PRA, and EIOPA SII L1 provide details about how to handle AI/ML model governance.

SR11-7[2]

This is a regulatory standard created by the U.S. Federal Reserve Bank that guides model risk management. This regulatory law provides guidelines to adhere to it. They are as follows:

- Model development, implementation, and use

- Model validation (conceptual soundness, monitoring, and analysis)

- Governance, policies, and controls

OSFI E23[3]

This is a risk management framework created by the Canadian government for banks, bank holdings, federally regulated trusts, and loan companies. This framework provides broader details of the following:

- Appropriate and commensurate governance system over model usage

- Policies and processes around model selection and development

- Continually assess model performance and suitability

- Audit functions to independently assess the model risk management governance and compliance framework

---

[2] *https://kristasoft.com/how-to-comply-with-sr-11-7-guidance-on-model-risk-management/*

[3] *https://www.osfi-bsif.gc.ca/Eng/fi-if/rg-ro/gdn-ort/gl-ld/Pages/e23.aspx*

PRA (Prudential Regulation Authority)[4]

This regulation is defined for banks, building societies, credit unions, insurers, and major investment firms. In July 2022, the UK government published the policy and approach to regulating AI/ML models. A few specific characteristics are as follows:

- Ensure that AI/ML is used safely.

- Ensure that AI/ML is technically secure and functions as designed.

- Make sure AI/ML is appropriately transparent and explainable.

- Consider fairness in AI/ML.

- Define legal person responsibility for AI/ML governance.

EIOPA SII L1 (European Insurance and Occupational Pension Authority)[5]

This is an independent European Union authority responsible for enhancing supervisory convergence and stability of the financial system within the insurance sector. The insurance sectors rely on AI/ML for various functions, such as underwriting, claims management, and pricing.

There are various compliance items and regulations in place for insurance sectors, as follows:

- Capital adequacy

- Transparency and disclosure

---

[4] *https://www.bankofengland.co.uk/prudential-regulation/ publication/2022/october/artificial-intelligence*
[5] *https://www.eiopa.europa.eu/system/files/2021-06/eiopa-ai-governance- principles-june-2021.pdf*

- Clear roles and responsibilities, oversight, and accountability

- Comply with GDPR.

Roles and Responsibilities

Organizations need to assess the skill requirements and responsibilities of each function. These roles are involved in different stages of the model lifecycle. Each role has specific tasks to comply with policies, statutory requirements, and codes of ethics to ensure successful implementation and management of AI/ML systems.

A few critical roles required in the datafication are as follows:

- **Data Engineers**: Engineering of data injection, collection and storage of data, and collaboration with other roles like data scientist and analyst

- **Data Scientist**: Develop the AI/ML models to analyze the patterns, trends, and insights. Collaborate with ML engineers, data engineers, and domain experts to improve the model's performance.

- **ML Engineer**: Develop ML models and maintain them in ML Pipelines (as explained in Chapter 8, "Datafication Engineering"), collaborate with data scientists, domain experts, and software engineers to refine the model.

- **Domain SME**: Translate business requirements into the AI/ML model questionnaire and evaluate the output based on the questionnaire.

- **Ethics and Compliance Officer**: Ensure that models comply with the laws, regulations, and ethical guidance.

- **Chief Data Officer**: Guide the entire team to follow the guidelines set by the organization.

Monitoring and Reporting

Datafication continuously evolves. To monitor data and models requires a dynamic approach. In the monitoring capabilities, you need to include performance management, reliability, detection of anomalies, transparency, and reporting and dashboarding.

In performance management, the monitoring tool collects metrics of the model runtime and detects anomalies to ensure the delivery of accurate and reliable results.

In drift management, monitoring the input data and model predictions helps to identify shifts in data distribution and model behavior over time.

In bias management, monitoring fairness measures helps identify and address potential biases in the model.

To monitor the preceding capabilities, you must choose appropriate tools, such as TensorFlow, MLFlow, Dynatrace, and so on.

Datafication Governance Guidelines and Principles

Datafication governance is based on guidelines and principles, frameworks, metrics, and processes that enterprises use to ensure responsible development, deployment, and management of data and AI/ML systems. To achieve overall governance, the datafication must consider the following critical principles and guidelines:

- The governance activities need to be incorporated into the development, deployment method for software, use of data for AI and ML models, and risk management.

- Successful governance starts with the organization's vision and commitment from leadership. Datafication activities such as data and AI/ML systems are guided by a strategy aligned with the organization's strategy.

- Governance is a shared responsibility between data stewards, data scientists, and other related professionals.

- The organization must be open about using AI and ML technologies and be transparent with details about how these systems work, including data sources, algorithms, etc.

- Datafication systems should respect individuals' privacy rights and comply with relevant data protection regulations. Organizations must ensure that data is stored and shared securely and ethically.

- Datafication must be aligned with the "design for security" principle to ensure that AI/ML models do not pose risks to people, property, or the environment.

- Datafication development must ensure transparency across all relevant stakeholders.

- Ensure the model's predictions are reasonably accurate and reliable and minimize the bias.

Ethical and Legal Aspects

In datafication, you must consider ethical principles and legal requirements for model development, training, validation and deployment, and operationalizing these models.

From data privacy and security, you need to consider both ethical and legal perspectives. Ethically, you must respect individuals' right to control their personal information, maintain confidentiality, and avoid data misuse. For the legal aspect, you need to adhere to the country-specific laws and regulations, such as GDPR from European Union, DPDPB from India, and so on.

From a bias perspective, an ethical model should not discriminate against any individuals based on race, gender, age, and so forth, and legally you need to comply with the local laws of each country.

From a transparency perspective, models must be transparent in decision making to ensure stakeholders understand and trust the outcomes.

From a human and machine perspective, models should be designed to augment human decision making rather than replace it. Ensure humans retain control over models.

Datafication Governance Action Framework

The action framework, as shown in Figure 9-2, outlines the specific steps and strategies that stakeholders can take to ensure responsible and ethical development and use of data and AI/ML models. The action framework illustrates the various components in a governance framework that interact and execute responsibilities.

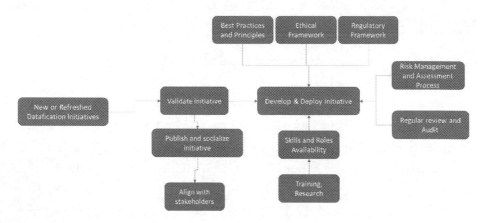

Figure 9-2. *Governance framework action*

The governance framework starts with any new or refreshed datafication initiative by an organization. Upon initiation, it has to publish and be validated with relevant stakeholders, including privacy officers, legal system, society, and governments.

The next step of the governance framework is to develop the initiative using the data analysis framework described in Chapter 5, "Data Analysis." During development, you must apply best practices and principles, the ethical framework of an organization and society, and the regulatory framework of each specific country. You need proper skills and roles to develop and deploy entire initiatives.

During development and deployment, you must conduct regular risk assessments and auditing of models by the governance board or a third-party risk accessor.

Datafication Governance Challenges

Datafication governance is complex and poses many challenges. Addressing these challenges requires a multi-disciplinary approach that involves various stakeholders, including legal experts, chief privacy officers, policy makers, and broader society. It is essential to develop governance frameworks that are agile and globally coordinated. A few challenges in governance are as follows:

- The AI and ML technologies are new and progressing rapidly, making organizations face difficulty in framing the governance framework to keep it up to date. The governance framework must be agile enough to respond to the changes to mitigate this advancement.

- There is a lack of clear standards and regulations available across the globe for the development and use of models. A few mature countries are creating a regulatory framework but are still not in the final state. This makes it difficult for organizations to know what is expected of them.

- Skill level is among the most significant concerns in AI/ML system development and deployment. This makes it challenging to regulate and govern these technologies effectively. Training, research, and education are essential to improve the skill set and resources.

- Privacy and transparency are among the most significant concerns because AI/ML models require collecting personal information. Only a few mature economies have privacy laws, and the rest of the world has no rules framed. These frameworks and standards should address transparency, accountability, bias, and discrimination.

Summary

In this chapter, we have explored the importance of risk and the governance framework, including the rapid pace of technological change, the lack of clear standards, ethical concerns, and the need for global governance. We have also discussed best practices and principles associated with governance. By following these best practices, you can ensure that AI/ML models are developed and used to benefit society and respect human rights and dignity.

In the end, we have discussed the challenges posed by governance.

In this chapter, I am providing a broad introduction to governance and its framework. This may only cover some of it, but you'll have an understanding of governance and can explore more details on your own.

CHAPTER 10

Datafication Security

In recent years, the proliferation of data and increasing reliance on digital and artificial intelligence (AI) / machine learning (ML) models has led to the transformation of various aspects of our lives into data that can be analyzed, processed, and used to make predictions.

AI/ML systems rely on the large amount of data to train algorithms, and these data sets contain sensitive information about users, consumers, and so on. In addition, the AI/ML models are vulnerable to various attacks, such as adversarial attacks, data poisoning, and model inversion attacks, which can compromise the integrity of the output.

As a result of these myriad vulnerabilities and the advancement of technologies, security has become a critical concern for organizations. All organizations must comply with the regulations drafted by multiple governments across the globe.

Datafication security is a vast topic. To cover everything, you need to read various security-related books on data and AI/ML, but in this chapter, I will provide essential details of datafication security and a security framework.

Introduction to Datafication Security

Datafication converts various aspects of human life into data and prediction. Its use requires security policies to ensure that the data is collected, stored, and used both ethically and responsibly.

© Shivakumar R. Goniwada 2023
S. R. Goniwada, *Introduction to Datafication*,
https://doi.org/10.1007/978-1-4842-9496-3_10

According to a survey conducted by a leading consulting company, 62 percent of AI and ML adopters see cybersecurity risks as a primary or extreme concern, and the rest of the survey participants mentioned that they are prepared to address those risks.

Datafication security includes the planning, development, and execution of security policies and procedures to provide proper authentication, authorization, access, cybersecurity, and third-party access. Datafication security requirements may be different across industries and countries. In a nutshell, data security aims to protect information assets such as data sets and models via privacy, regulations, and confidentiality.

The security policies in datafication are there to ensure that authorized people can use and update data in the right way. The goals of the datafication security are as follows:

- Understand and comply with all relevant regulations and policies for privacy, protection, and confidentiality.

- Make sure privacy and confidentiality are protected from unauthorized access or disclosure. This multi-level authorization includes privileged access control (PAM), data encryption, and anonymization techniques.

- Ensure data validation and verification process is in place to control adversarial attacks.

- Ensure your models are trained and validated against unbiased and diverse data assets.

- Conduct regular vulnerability assessments on code by using various tools to mitigate the OWASP Top 10 vulnerabilities and human errors.

Datafication Security Framework

Effective datafication security policies should be comprehensive to ensure that the right people can correctly manage the data and AI/ML models. Figure 10-1 shows that understanding and complying with security policies, regulations, organizations' concerns, governance and compliance, and business access needs is essential.

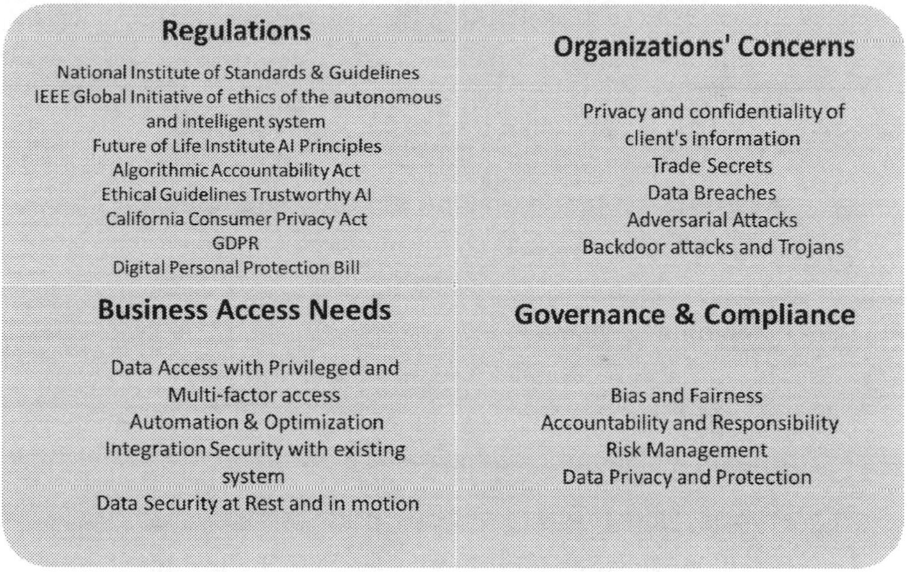

Regulations	Organizations' Concerns
National Institute of Standards & Guidelines IEEE Global Initiative of ethics of the autonomous and intelligent system Future of Life Institute AI Principles Algorithmic Accountability Act Ethical Guidelines Trustworthy AI California Consumer Privacy Act GDPR Digital Personal Protection Bill	Privacy and confidentiality of client's information Trade Secrets Data Breaches Adversarial Attacks Backdoor attacks and Trojans
Business Access Needs	**Governance & Compliance**
Data Access with Privileged and Multi-factor access Automation & Optimization Integration Security with existing system Data Security at Rest and in motion	Bias and Fairness Accountability and Responsibility Risk Management Data Privacy and Protection

***Figure 10-1.** Datafication security framework*

By implementing these central policies and procedures, organizations can reduce the risk of security and compliance incidents and protect datafication programs.

Regulations

Various regulations are available globally, but these are evolving as AI/ML technologies evolve; regulations and guidelines are being developed to ensure their ethical and responsible use. Everyone who initiates the AI/ML

programs must understand and comply with these regulations to ensure
that the use of AI/ML aligns with legal and ethical standards. The globally
available regulations are as follows:

- **National Institute of Standards and Guidelines for
  AI Standards (NIST)**: This framework provides a set of
  standards and guidelines to support the development
  and use of AI/ML technologies. This framework covers
  critical areas such as data quality, risk management,
  explainability and transparency, robustness and
  security, privacy, and interoperability. You can find
  more details on the NIST[1] website.

- **IEEE Global Initiative of Ethics of the Autonomous
  and Intelligent System**: This framework is for the
  ethical and responsible development and use of
  autonomous and intelligent systems, AI, and robotics.
  This framework provides a set of ethical principles,
  technical standards, education awareness, and so forth.
  You can find more details on the IEEE[2] website.

- **Future of Life Institute AI Principles**: This framework
  provides guidelines for the ethical development and
  use of AI technologies. You can use these principles
  to ensure that AI/ML technologies are aligned with
  human values and are transparent, unbiased, fair, and
  secure. You can find more details on the Future of Life[3]
  website.

---

[1] *https://www.nist.gov/artificial-intelligence*

[2] *https://standards.ieee.org/industry-connections/ec/
autonomous-systems/*

[3] *https://futureoflife.org/cause-area/artificial-intelligence/*

- **Algorithmic Accountability Act**: U.S. federal law
  proposes this act (some congressmen proposes, it
  become law if it passes it). It is meant to promote
  transparency and accountability in using AI/ML
  systems and to prevent discrimination and bias. This
  law covers vital capabilities such as discriminatory
  impact assessment, notice and consent from each
  person before collecting the data, mitigation details for
  any risks, and civil rights protection. (This law is still in
  the draft stage).

- **Ethical Guidelines Trustworthy AI**: This framework
  was created by an expert group from the European
  Commission for ethically responsible development.
  The guidelines of this framework emphasize the
  importance of developing AI/ML technologies that
  augment human decision making and agency rather
  than replacing them and the importance of developing
  AI/ML technologies that are technically robust and
  safe. This framework also provides for privacy and
  data governance and societal and environmental
  well-being. With these guidelines, you can ensure that
  AI/ML technologies are aligned with human values,
  are technically robust and safe, and are respectful of
  privacy and data governance.

- **California Consumer Privacy Act**: This was created
  in 2020 in California. This act provides comprehensive
  privacy regulations that give consumers the right to
  know what personal information is being collected,
  who has access to individual personal information, and
  so forth.

- **Digital Personal Protection Bill**: This bill was created by the Indian government in 2022 to protect and store personal data. The critical factors of this law are to protect personal data, require consent from individuals to collect the data, provide the right to get information about personal data, and ensure the right of grievance redressal.

- **GDPR (General Data Protection and Regulation)**: The European Union developed this regulation in 2018. The key rights of individuals under this regulation are the right to access personal data, the right to rectify personal information, the right to delete personal information, the right to restrict processing, the right to data portability, and the right to object.

Organization Concerns

Organizations have several security concerns in terms of datafication. They are privacy, trade secrets, data breaches, security at rest, adversarial attacks, model stealling and reverse engineering, inside threats, data poisoning and manipulation, and so on.

For privacy, you need to adhere to regulatory frameworks and laws, as mentioned in the previous section, for all of your datafication programs.

Trade secrets play an essential role in developing and protecting data and AI/ML models. Here, trade secrets refer to the confidential information kept secret within an organization to maintain competitive advantage, protect innovation, and so on. To protect trade secrets, you must implement various steps, including stopping unauthorized access, signing NDAs (non-disclosure agreements), and monitoring these assets.

A data breach is about publicly breaching sensitive data, models, and exposed non-disclosed information. For example, recently, the Twitter code has been shared publicly. This is a violation of trade secrets and a data breach. Data breaches in datafication can occur for several reasons: insider threats, compromise of cybersecurity, lack of data protection measures, unauthorized access, etc. To mitigate all these threats, you need to implement firewalls, encryption, access controls, audit and risk assessment, breach response plans, and so forth.

The adversarial attack is a technique that is used to manipulate datafication models to produce misleading or biased results. There are various adversarial attacks, such as injecting malicious data into the training data so it is used to build machine learning models, manipulating the input data to create biased output, reverse engineering your model, and more. You need to implement security measures such as access control, encryption, and monitoring tools. Along with these techniques, you need to conduct regular risk assessments and audits.

Backdoor attacks and trojans are malware designed to gain unauthorized access to your organization's network or computer system. Hackers usually use this method. To minimize these attacks, you need to consider implementing security measures such as perimeter security, regular risk and audit management, fake attacks to check vulnerability, stress tests, and monitoring network and individual systems.

Governance and Compliance

An effective governance structure is necessary to ensure that organizations have a clear direction for datafication security. The datafication models are complex and make it challenging to identify potential security risks. It is essential to have adequate security policies and procedures in place to manage the AI/ML models.

As shown in Figure 10-2, a security governance structure must coordinate various stakeholders, have security practices and guidelines defined, offer oversight and escalation for any severity, and balance concerns across different teams.

Establish forums for security awareness, transparency, and strategic guidance among business, technology and security team.

Embed security practices into day-to-day AI/ML management

Achieve balanced oversight with Business Unit representation.

Provide clear point of escalation and status elevation to the Executive Steering Committee and Board of Directors.

Formalize linkages between Security, Risk Management, and other key functions.

Vest major security decisions with the "authority" of leadership across an organization.

Figure 10-2. *Security governance structure*

Within the structure, the critical elements are as follows:

- **Policies and Procedures**: The policies and procedures define the security requirements for the organization, such as encryption, stopping unauthorized access, data protection, vulnerability check, etc.

- **Risk Management and Risk Assessment**: This is an essential element for any datafication program, which involves assessing the security risks and conducting a regular assessment to ensure everything is managed as expected.

- **Incident Response**: It is tough to say your system is 100 percent safe and secure. Every day there will be a new malware attack on each organization, so every organization must have an incident response procedure.

- **Compliance and Governance**: Make sure that all datafication programs follow the compliance and standards as defined in the regulation section.

- **Ethics, Privacy, and Bias**: In the governance structure, you need to have clear policies and procedures to ensure that the AI/ML models are developed, trained, and used in a way that is ethical and aligned with the country's regulations.

By including these elements in your governance, it ensures that AI/ML programs are managed in a way that is secure and aligned with security objectives.

Business Access Needs

Business access in AI/ML models requires a comprehensive approach that includes data and system access, automation and optimization, integration of access to the various systems, like API access, and data security both at rest and in motion.

The business access requirements are categorized into three groups—access, audit, and authorization—as follows:

- Access is enabled for all the datafication models. There are various methods to be considered, such as classic user ID and password, single sign-on, and access through unique identification (SSN, Aadhar in India, Emirates ID in UAE, etc.).

- Auditing datafication to review security actions and access activity to ensure compliance with regulations and conformance to policy and standards. This auditing helps you to detect unauthorized access and breaches into your data and models. For effective auditing, you need to consider the entire monitored activity of the accessor that could indicate potential security risks and incidents. While auditing AI/ML systems, you must remember regulatory details because AI/ML models are used in various prediction processes.

- Authorization grants individuals privileged access to specific users based on the roles and groups for data and models. It is essential to have robust authorization management in place that defines who is allowed access to the data and models. When working on the authorization process, you must consider role-based access controls, data classifications, and classifying data based on sensitivity risks. Data classification ensures that sensitivity is protected and data is only accessible to authorized users.

Datafication Security Measures

The datafication security measures are the steps taken to protect AI/ML systems—from application, integration, and infrastructure perspectives—against inside and outside threats. The following are the key security measures.

Encryption

Encryption translates plain text into complex coded language to hide privileged information and can only be read or processed by a decryption key. This includes encrypting data at rest and in transit. Before applying the encryption to data, consider doing data classification first. The classification helps you to identify sensitive fields to take a call, either encryption or masking.

There are a few main methods of encryption available: hash, symmetric, and asymmetric.

It would be best if you kept an eye on key management, rotation of keys, and revocation. These keys should be stored in critical management software like Hashicorp vault.

Data Masking

Data masking is the process of modifying sensitive data in such a way that it becomes unreadable by unauthorized users while still retaining its original format.

There are various masking methods: character substitution, word substitution, truncation, and tokenization. The choice of masking depends on the specific needs of the AI/ML models and the level of security required.

Penetration Testing

Penetration testing tests the security of a computer system, network, or AI/ML systems by simulating an attack from a malicious actor. Penetration testing aims to identify vulnerabilities and weaknesses in system security before attackers exploit them.

Data Security Restrictions

There are two types of data security restrictions: level of confidentiality and regulation related to data.

> **Confidentiality Level**: This restriction protects sensitive information from unauthorized access. To ensure confidentiality, several measures can be taken, such as encryption, masking, data classification, and infrastructure and physical security.

Regulation Related to Data: Regulatory categories are assigned based on the regulation laws, which can be GDPR, HIPPA, California Consumer Privacy Act, Digital Personal Protection Bill, etc.

Summary

Every corner of the globe has increased its reliance on AI/ML technologies to solve various problems, but these technologies have also introduced new security challenges. AI/ML security is a critical concern for organizations, and they must protect the integrity and confidentiality of their data and model.

To ensure the security of AI/ML systems, organizations must implement the highest standard of security measures, such as secure data management and secure model development.

In this chapter, I introduced the various aspects of the security mechanism for datafication. You can use this chapter as an intermediary and seek security-related books for your advanced knowledge.

Index

A

Access testing, 229
Adversarial attack, 229, 251, 256, 257
Adversarial testing, 229, 230
Agile methodologies, 231, 232
AI and ML development
    cognitive models, 219
    development workplace, 234
    experiment management, 235
    feature engineering, 234
    feature store, 235
    getting cleansed data
        correlation-based feature analysis, 221
        feature selection and engineering, 221
        guidelines, 220–221
        PCA, 222
        recursive elimination feature selection, 222
    identify objective/problem, 217
    ML models, 218
    model training, 235
    simulation models, 219
    statistical models, 218

AI and ML engineering, 215, 218, 219, 222
    DataOps, 231–234
    MLOps, 233–235
    model training and optimization, 222–223
    steps, development
        challenging the result, 217
        data collection, 216
        data processing, 216
        deployment, 217
        identify source, 216
        interpretation of result, 217
        model selection, 216
        problem definition, 216
        statistical prediction modcling, 216
        train the model, 217

AI and ML testing, 223
    integration testing, 225
    non-functional testing, 225–226
    performance testing, 227
        availability and scalability, 227
        benchmark testing, 227
        data size, 227
        fitness testing, 228

Printed in the United States
by Baker & Taylor Publisher Services